T0136723

# THE GREATEST ADVENTURE

KOSMOS

A series exploring our expanding knowledge of the cosmos through science and technology and investigating historical, contemporary and future developments as well as providing guidance for all those interested in astronomy.

Series Editor: Peter Morris

Already published:

*Asteroids*  Clifford J. Cunningham
*The Greatest Adventure*  Colin Burgess
*Jupiter*  William Sheehan and Thomas Hockey
*Mars*  Stephen James O'Meara
*Mercury*  William Sheehan
*The Moon*  Bill Leatherbarrow
*Saturn*  William Sheehan
*The Sun*  Leon Golub and Jay M. Pasachoff

# THE
# GREATEST
# ADVENTURE

*A History of
Human Space Exploration*

## COLIN BURGESS

REAKTION BOOKS

To our Fallen Heroes: the crews of Apollo 1 (AS-204), Soyuz-1, Soyuz-11, *Challenger* and *Columbia*

Published by Reaktion Books Ltd
Unit 32, Waterside
44–48 Wharf Road
London N1 7UX, UK
www.reaktionbooks.co.uk

First published 2021
Copyright © Colin Burgess 2021

Printed and bound in Great Britain by Bell & Bain, Glasgow

A catalogue record for this book is available from the British Library

ISBN 978 1 78914 460 4

# CONTENTS

# PROLOGUE

Just as we continue to do today, our prehistoric ancestors would have gazed in profound wonderment at the rising silvery Moon. They could not fathom its purpose; all they knew was that its brilliant fullness gave a welcome light in the darkness and that it might exude some mystical, magical powers.

Few people understood that this and other venerated astronomical bodies they could see in the night sky might be solid, independent worlds; instead, they became a visible source of myths and fantasies, preserved and immortalized through successive generations and across many cultures. These myths have concerned such legendary Roman and Greek heroes as Selene, Diana, Astarte, Luna and Daedalus, as well as Soma, a Hindu god of the moon, and the Aztec sun and moon gods Nanahuatzin and Tecciztecatl.

Astronomy, the study of our Moon, the planets, stars and other celestial objects, is widely regarded as our earliest science. Once astronomers and philosophers began to comprehend our planet's actual position in the universe, those myths of old mostly gave way to later theoreticians musing about ways to somehow fly to our closest celestial neighbour.

Speculation about the Moon intensified when Galileo used a crude refracting telescope to gaze not only upon our cratered Moon but beyond, to the moons of Jupiter. Later, circa 1609, fellow astronomer Johannes Kepler wrote a book called *The Dream*, a popular fantasy in which demons begin spiriting people away to the Moon.

In his classic 1865 novel *From the Earth to the Moon*, French science fiction writer Jules Verne prophetically told of a lunar flight successfully carried out by three intrepid explorers. Incredibly, their fictional journey would bear many striking parallels with the first lunar landing by human beings more than a century later. The explosive propulsion device used to launch Verne's three-man aluminium spacecraft is called Columbiad, while the Apollo 11 spacecraft was named *Columbia*. Remarkably, the production and use of aluminium in manufacturing was relatively undeveloped at the time of Verne's book. The Columbiad craft is propelled on its journey to the Moon from Florida, not too far from the present-day Kennedy Space Center, and, just like Apollo 11 in 1969, it splashed down in the Pacific Ocean for recovery by naval vessels. There are many other startling similarities: one of Verne's lunar travellers has the surname Ardan, while Buzz Aldrin was a crew member on Apollo 11; furthermore, Neil Armstrong's middle name was Alden. Another adventurer is named Nicholls – a near anagram of the surname of command module pilot Michael Collins. Verne also predicted the use of the spacecraft propulsion device known as a retrorocket, many decades before they were actually developed. Unlike Apollo 11, however, expedition leader Barbicane's crew carry along some non-human domestic passengers, as seen in contemporary woodcut illustrations featuring a pair of cockerels and two small dogs, one of which has the intriguing name of Satellite. Verne may rightfully be regarded as a true doyen of science fiction, but even he could not have envisaged the rich and dramatic history that would unfold a century beyond the publication of his wonderfully prescient book.

Along with advancements in science and technology came the discouraging knowledge that we are firmly in the grip of gravity and that there is no simple method of breaking free from our planet. During the late nineteenth century there was a growing realization that only mighty rockets would be capable of defying the immutable laws of gravity, carrying vehicles beyond the atmosphere and into space. In a paper published in 1903, Russian science teacher Konstantin Tsiolkovsky foresaw the development and construction of rockets fuelled by liquid propellants, including liquid hydrogen and liquid oxygen. This would later be the combination employed in launching the first successful rockets and, decades on, America's fleet of space shuttles. An avid reader of science fiction, Tsiolkovsky calculated that Verne's description of using the massive Columbiad space cannon to propel his explorers to the Moon was flawed, as the massive acceleration forces created would crush any occupants to a bloody pulp inside their capsule. He relied instead on Newton's third law of motion: that for every action, there is an equal and opposite reaction – the very basics of rocketry. Hailed as a true genius by the Soviets, Tsiolkovsky would come to be recognized as the father of astronautics.

There were others undertaking similar pursuits in rocketry around this time. Foremost among those in the United States was the physicist Robert H. Goddard, who began building and launching tiny rockets in 1914, much to the wary bemusement of people residing in Auburn, Massachusetts, who dismissed his quaint rockets as the work of a harmless eccentric. Undeterred, Goddard continued with his work and successfully launched the first liquid-fuelled rocket to an altitude of 56 m (184 ft) on 16 March 1926. Over the next fifteen years, Goddard and his helpers managed to launch no fewer than 34 rockets, eventually reaching a maximum altitude of 2.6 km (8,500 ft) and speeds close to 885 km/h (550 mph). Despite receiving very little financial or public support for his research work and test firings, Goddard would finally become known as one of the founding fathers of modern rocketry. In 1904, he told his high school graduating class, 'It is difficult to say what is impossible, for the dream of yesterday is the hope of today – and the reality of tomorrow.'

Other visionaries included physicist and engineer Hermann Oberth, who published his 92-page work *Die Rakete zu den Planetenräumen* (The Rocket into Interplanetary Space) in Germany in June 1923. Six years later, he expanded on this dissertation in a full-length book, *Wege zur Raumschiffahrt* (Ways to Space Flight). In the autumn of that year, 1929, he conducted a static firing of his first liquid-fuelled rocket motor, christened the *Kegeldüse* (meaning 'cone nozzle'). The rocket's engine was developed for him by fellow engineer Klaus Riedel in a small workshop located on proving grounds provided by Berlin's Reichs Institute for Chemical Technology. The *Kegeldüse* fired and ran briefly. It was a start.

One of those assisting Oberth and Riedel and observing the firings was an enthusiastic eighteen-year-old German rocketry student named Wernher von Braun. 'Our equipment was elementary, and our ignition system was perilous,' von Braun later recalled. 'Klaus Riedel would toss a flaming gasoline-soaked rag over the gas-spitting motor, and then duck for cover before Oberth opened the fuel valves and it started with a roar.'[1]

Years later, during the Second World War, von Braun was involved in developing supersonic rockets for the German war effort. In the latter war years, the Nazi regime provided funding that made possible the construction and firing of Hitler's weapons of destruction, the v-1 and v-2. In early 1945, as the end of the war seemed inevitable, invading Russian forces began closing in on the missile firing site at Peenemünde, located on the Baltic Sea island of Usedom, seeking the valuable spoils of war. When these forces came within 160 km (100 mi.) of Peenemünde, von Braun assembled his rocket team and declared they would be far better off surrendering to American forces than face the renowned brutality of the Soviets.

As the first part of von Braun's plan, he was able to falsify travel documents, and the group of planners made their way to the Mittelwerk underground factory, located beneath a hill in Thuringia's Harz mountain range. Here, forced labour from concentration camps had been used in the manufacture of the deadly v-1 and v-2 rockets. On arrival, von Braun's rocket team resumed work on the missile programme, while keeping a cautious watch on the progress of the war. Knowing that the ss would probably want to destroy any rocket blueprints to prevent them falling into Allied hands, von Braun had them secretly sealed in a mine shaft for possible exploitation in seeking a surrender deal.

Rocket engineer and designer Wernher von Braun (with broken left arm) surrenders to u.s. forces. To his right (in hat) is Major General Walter Dornberger, head of Hitler's v-2 programme.

In March, von Braun suffered a broken arm when his driver fell asleep at the wheel of a car and crashed. The rocket designer was hospitalized and his left arm placed in a plaster cast. The following month, as Allied troops advanced ever deeper into Germany, von Braun managed to convince his superiors to move a large number of his engineering team to more secure territory in Austria. His strategy worked perfectly; on 2 May 1945, his brother Magnus, a fellow rocket engineer, surrendered von Braun's team to members of General Patton's 44th Infantry Division.

Once the surrender formalities had been taken care of, and the Allies understood the importance of their prisoners, Wernher von Braun revealed where the blueprint documents could be found and retrieved, along with crucial sections of one hundred v-2 rockets. After the war, von Braun and a vast cadre of captured or surrendered rocket scientists, technicians, engineers and others who had designed or worked on these rockets were transported to the United States. In a speech given in 1962, von Braun recalled their first years studying and working on American soil:

> The next five year period came at Fort Bliss, Texas, from September 1945 to April 1950, where we worked for the u.s. Army. About 120 handpicked members of the v-2 team were gradually supplemented by about 400 civilians and soldiers of the u.s. Army Ordnance Corps. Our first year here was a period of adjustment and professional frustration. Distrusted aliens in a desolate region of a foreign land, for the first time we had no assigned project, no real task. Nobody seemed to be much interested in work that smelled of weapons, now that the war was over. And space flight was a word bordering on the ridiculous.
>
> We spent the time in study and teaching, and assisted with the v-2 evaluation firings at White Sands, New Mexico. In addition to rocketry, the German-born members of our team studied the American language, American Government, and the American way of life. These were our years of wandering in the wilderness.[2]

In 1950 the u.s. Army decided to relocate von Braun's Peenemünde group, then composed of some 1,600 personnel, from Fort Bliss to the Redstone Arsenal in Huntsville, Alabama, in order to create a foundation rocketry team. Army chiefs were becoming increasingly aware that the Soviet Union had made significant gains in missile technology since the war and the Department of Defense was keen to speed up progress on the development of intercontinental ballistic missile (ICBM) and intermediate-range ballistic missile (IRBM) weapons systems as a national priority.

By 1955, it was becoming increasingly obvious that the Soviets were well ahead in missile technology, and a serious 'missile gap' was causing profound concern. With this in mind, the U.S. Army was directed to cooperate with the U.S. Navy in developing a land- and sea-based IRBM. A Joint Army–Navy Ballistic Missile Committee was formed, and on 1 February 1956 the Army Ballistic Missile Agency (ABMA) was established under Major General John B. Medaris with a mandate to develop the IRBM, a rocket that came to be known as the Jupiter. Von Braun was rapidly becoming the person most instrumental not just in the pioneering days of the American space programme, but in placing the first human beings on another heavenly body – the Moon.

By this time, both the United States and the Soviet Union had been conducting biological suborbital space flights, launching primates and canines respectively on short ballistic flights to determine their reaction to the forces associated with launch and brief periods of weightlessness. Then, on 4 October 1957, the Soviet Union launched the world's first satellite into space, catching America off-guard and fearful of a Cold War enemy who could now potentially rain nuclear weapons on the United States from space.

The following month a second satellite was sent into orbit, this time carrying a small dog. With the U.S. military now fully occupied in the development of ICBMs and IRBMs and supersonic, high-altitude aircraft, together with persistent rumours that the Soviet Union was preparing to send the first man into space, President Dwight Eisenhower and his advisers decided there was an urgent need to establish a civilian space agency to develop a manned space flight capability.

Since 1915, the National Advisory Committee for Aeronautics (NACA) had been working on problems associated with aircraft flying faster and higher, incrementally extending its research to high-altitude and space flight. In April 1958, Congress took the first steps to set up the new civilian space agency, and President Eisenhower signed the National Aeronautics and Space Act three months later, on 29 July. A further three months on, the National Aeronautics and Space Administration (NASA) officially came into being, headed by the agency's first administrator, T. Keith Glennan. On 1 October, NASA began operations by fully absorbing NACA, its 8,000 employees and five research centres, and its $100 million annual budget.

What was later to become known as the Space Race to the Moon truly began on 12 April 1961 with the shock announcement that the Soviet Union had launched a man into space, returning to a safe landing after completing a single Earth orbit. His name was Yuri Alekseyevich Gagarin, a lieutenant in the Soviet Air Forces, promoted to the rank of major during his epic mission.

Gagarin's history-making flight came just three weeks before America, via NASA, was preparing to send Navy commander Alan Shepard on a suborbital space flight, which by comparison looked technologically inferior to that of Gagarin. It was a tough time for everyone at NASA, and newly elected U.S. president John F. Kennedy must surely have gritted his teeth when he sent a congratulatory cable to Soviet premier Nikita Khrushchev. In his message, he said that America shared with the Soviets 'their satisfaction for the safe flight of the astronaut in man's first venture into space . . . It is my sincere desire that in the continuing quest for knowledge of outer space our nations can work together to obtain the greatest benefit to mankind.'³

At his later press conference, President Kennedy conceded that it was going to take some time for the United States to catch up with the Soviet ability to launch heavy payloads into space. Furthermore, he expressed his hope that America would make significant progress 'in other areas where we can be first and which will bring perhaps more long-range benefits to mankind. But here, we're behind.'⁴

The president knew he had to make some bold moves in order to placate the anxious American people, who were openly questioning why the United States had been allowed to slip behind the gloating Soviets in space achievements. Resolving to address the situation, he called together many of his closest political advisers and members of the NASA hierarchy, asking for their advice, opinions and suggestions, while telling them that a bold plan of action was required. Kennedy also needed to overcome the embarrassing fallout from the politically calamitous and costly Bay of Pigs invasion of Cuba, which, for many citizens, had portrayed him in the very infancy of his presidency as a poor and indecisive leader. He listened to his advisers and then made certain determinations.

On 5 May 1961, Alan Shepard rode a Redstone rocket on his sub-orbital mission aboard a Mercury spacecraft named *Freedom 7*, splashing down safely fifteen minutes after lift-off. America was ecstatic to finally have one of their own people in space, and the nation's jubilation convinced President Kennedy to make the trip across Washington from the White House on a warm spring day less than three weeks later. The date was 25 May 1961, four days short of his 44th birthday, and the president was about to deliver a speech outlining a monumental undertaking before a joint session of Congress.

In one of President Kennedy's most passionate, eloquent and oft-recalled speeches, he told Congress:

> I believe that this nation should commit itself to achieving the goal, before this decade is out, of landing a man on the Moon and returning him safely to the Earth. No single space project in this period will be more impressive to mankind, or more important for the long-range exploration of space. And none will be so difficult or expensive to accomplish.[5]

Committing the United States to a manned lunar landing was not only an audacious move but one involving considerable political risk. It was an undertaking given at a time when the nation's only manned space flight experience was a mere fifteen-minute suborbital lob earlier that month. Nevertheless, America's space programme, now with the promise of extensive funding to come, meant that everyone involved could look to the future, even though the eight-year time frame outlined by Kennedy was daunting and quietly perceived by many as highly optimistic. Regardless, preliminary work and planning began.

In the years to come, America's space programme would enjoy tremendous advances in technology, with NASA's astronauts accomplishing goals of increasing complexity, length and difficulty, despite some tragic setbacks. Although he never lived to see the first American astronauts walk on the Moon before the end of the 1960s, the Apollo programme that placed them on the lunar surface will forever exemplify the spirited character and commitment of Kennedy's presidency.

President John F. Kennedy addresses a joint session of u.s. Congress, 25 May 1961, pledging the nation to a lunar landing goal by the end of the decade.

IN APRIL 2021, the world commemorated the sixtieth anniversary of the first human space flight by Soviet cosmonaut Yuri Gagarin. When that anniversary rolled around, somewhere in the vicinity of 550 people had been launched into space, slightly less than two-thirds of them Americans, and a number that disproportionately includes fewer than seventy women.

To date, there have been eighteen astronaut and cosmonaut fatalities during space flights, including fourteen people lost in the *Challenger* and *Columbia* space shuttle tragedies. Several astronauts have also died while training for space missions, including the three Apollo crew members who perished in a ferocious launch pad fire in January 1967. Other fatalities occurred during space flight-related activities, including 24-year-old Valentin Bondarenko, a member of the Soviet Union's first cosmonaut team, who died in a test-chamber fire fuelled by a pure oxygen environment in March 1961. Despite these tragedies, all those participating knew and accepted the hazards involved in flying into space, and mourned the loss of their colleagues. They realized that space exploration is a human imperative and that it would continue despite the losses.

In those six pioneering decades we have witnessed the origins of the Space Race, the first man and woman in space, the first spacewalks, the first rendezvous and docking in orbit, the first humans setting foot on the Moon, numerous protracted science missions, and the occupation of orbiting Russian, American and Chinese space stations. The International Space Station (ISS) remains in orbit and has been under continuous occupancy since the first long-term residents arrived on 2 November 2000.

As we prepare for the next six decades of space travel, which will undoubtedly include fully commercial journeys by numerous national and private participants, whole new generations of space explorers will travel beyond low Earth orbit once again, on journeys to the Moon and Mars and many other exciting possibilities within our solar system.

Unlike those people from ancient and even more recent times, who could only gaze at the Moon and contemplate what it would be like to travel to that magical place and other worlds, we have collectively witnessed those fledgling steps out into the universe. Travelling further afield in space is our undeniable destiny, and given the spur of human curiosity to seek and explore, such aspirations are both beckoning and achievable.

# 1

# THE REALITY OF TOMORROW

Once considered inaccessible, Earth's jungles, seas, deserts and North and South Poles have all been conquered over the centuries through the enterprise and exertions of countless explorers. While still unexplored terrestrial corners of this great blue planet remain, there is another seductive frontier we are only just beginning to penetrate and understand, and it is located only a few kilometres from wherever we live. It is a vertical frontier, one far more challenging and aggressively hostile than any we have ever faced before. It is space.

Throughout history, humans have experienced a compulsion to go beyond what we have previously known. One such urge has always focused its attention on our Moon, that wondrous sphere in the night sky that has long held an irresistible fascination. First, though, we needed to know more about the limits and dangers of the atmosphere that engulfs and protects us.

In the middle years of the twentieth century, heroic balloonists ascended into the frigid regions of the upper atmosphere, where the huge, gas-filled spheres above their heads literally froze to the fragility of a light bulb. After the Second World War, competitive aviators pushed their supersonic steeds to the very limits of human and mechanical endurance, coaxing planes to achieve heights where no air passed over the wings to create lift. They would find themselves entering an unforgiving environment where any aircraft – no matter how powerful or aerodynamic – can suddenly surrender to the unyielding laws of high-altitude nothingness and begin tumbling out of control.

It was not just a matter of developing aircraft to conquer this tantalizing high frontier. Pressurized cabins, breathing masks and high-altitude suits – all had to be devised and manufactured, allowing pilots to survive in the near vacuum of the upper atmosphere. Barely 18 km (60,000 ft) above ground level, the air pressure is so weak that any pilot not wearing suitable protection and breathing apparatus would be deprived of oxygen and black out from the effects of hypoxia. Furthermore, the lack of a suitable pressure suit at extreme altitudes would cause their blood to bubble and fizz like uncorked champagne. One of the few positive results of the Second World War, as in any war, was the rapid advancement of technology, which in the post-war years would assist us in taking those first tentative steps out into the space frontier.

In 1945, with the collapse of Germany's Third Reich both inevitable and imminent, American and Soviet forces advanced through the vanquished nation, rapidly sweeping ever deeper into Germany's heartland. One of their prime objectives was to seize equipment and personnel from the once formidable and terrifying v-2 missile programme.

American forces had soon accumulated a vast collection of rockets and rocket parts, as well as dozens of captured or surrendered scientists and technicians. These men, their families, and all the components were soon dispatched overseas to a remote desert area of the American southwest known as White Sands, New Mexico. With his tremendous experience in the field of rocketry, from simple rockets through to Hitler's V-weapons, Wernher von Braun quickly established himself as the dominant figure in America's missile programme.

For their part, advancing Soviet troops did not fare quite as well, but still managed to seize a modest share of the v-2 booty and capture a number of German rocket technicians. The Soviet Union's missile development programme then continued with these new resources in the stark and forbidding outpost of Kapustin Yar, located on the empty steppes north of the Caspian Sea. In both nations the focus was on the development of missiles for potential military use, but for some of those involved there was also an unquenchable vision that more powerful variants of the rockets might one day be used to conquer space.

Our pathway to the stars began to unfold at some of the most inhospitable locations on the planet, with the work shrouded in a dark curtain of secrecy. It might be said, without too much exaggeration, that in the Soviet Union one man carried the vision of this future, along with the force of will to make it happen: Sergei Pavlovich Korolev.

*Opposite*: Loading alcohol fuel into a captured v-2 mounted on its portable gantry at the u.s. Army Ordnance Proving Ground, New Mexico.

No one burned with quite the same energy and urgency as Korolev, who for many years was known officially as the unseen, unnamed Chief Designer of the Soviet space programme. His name, and the enormous role he played in Soviet rocketry and astronautics, would not be revealed until after his untimely death in 1966.

Korolev was born in the Ukrainian town of Zhytomyr on 12 January 1907. In his youth he trained as an aeronautical engineer at the Kiev Polytechnic Institute, following which he became interested in rocket propulsion while attending Moscow University, later joining and becoming an active member of the Gruppa Izucheniya Reaktivnogo Dvizheniya (GIRD, or Group for the Study of Reactive Motion), a club dedicated to the development and testing of liquid fuel rockets. Before long, powers within Joseph Stalin's military recognized the impressive work being carried out by this civilian organization, and in 1933 it was taken over under his orders. The group then became known as Reaktivnyy Nauchno-issledovatelskiy Institut (RNII, or Research Institute for Jet Propulsion), the official centre for research and the development of missiles and rocket-powered gliders.

Korolev continued his work until June 1938, when he was swept up in one of Stalin's notorious purges during the barbaric period known as the Great Terror. Under spurious charges he was sentenced to ten

Sergei Pavlovich Korolev, the Soviet space programme's secret Chief Designer.

years' hard labour and sent to one of Stalin's brutal Gulags, where he would spend two years working on railways and ships and in gold mines. His previous work in rocketry eventually came to the notice of a former prisoner, famed aircraft designer Andrei Tupolev. While still regarded as a political prisoner, Korolev was permitted to work on Tupolev's design team.

Late in 1944, Korolev was authorized to form his own rocketry team and within three days had put together a proposal for a Soviet equivalent to Germany's v-2 missile. His rockets succeeded, although the expected range of the missiles was considerably less than that of the v-2. The following year he was sent to Germany in order to coordinate the recovery and assessment of v-2 parts and to recruit many of the experts who had worked on the destructive rockets in the latter years of the war.

On his return to the Soviet Union, Korolev was named as chief engineer in charge of a design team responsible for the development of a cosmetically modified, Soviet version of the v-2, which later became known as the R-1. Korolev would continue his pioneering work, and, following the death of Stalin in 1953, began enjoying the powerful support of the new Soviet premier, Nikita Khrushchev.

Several years later, the privations and working conditions he had endured while held in the Gulag camps contributed to Korolev's premature death in 1966, unfortunately at one of the most critical phases of the Soviet space programme. Following his death, Korolev's name no longer remained a closely kept state secret, and he quickly became a revered icon of Russian rocketry. It is a tribute to his work that both the rockets and spacecraft he helped design and develop, with some later modifications, are still flying today.[1]

## Dogs Riding Rockets

Under the watchful eye of Sergei Korolev, the Russians are believed to have launched the first of their captured German v-2 rockets on 18 October 1947, and their first modified prototype R-1 took to the skies from Kapustin Yar, on the banks of the Volga River, on 17 September the following year. Unfortunately the R-1 veered off course after launch, but a second test on 10 October proved more successful.

In order to conduct biological studies of organisms in space flight, the R-1B geophysical rocket was fitted with a recoverable payload section, which included a parachute system and a 'skirt' of six external drag brakes. During the separated capsule's descent it would free-fall for several minutes until it reached thicker air. The braking flaps then extended to steady and slow the capsule, after which a drogue chute would deploy, followed by a larger parachute.

It had already been decided that once larger nose cone capsules had been developed, the first creatures to occupy them would be small dogs. These would be 'recruited' from the streets of Moscow, which meant they were tough animals conditioned to a hard life, inured to hunger and freezing temperatures. Armed with a list of requirements, zoologists and other canine specialists were sent around Moscow to hunt for suitable candidates from among the common street type or mongrels. Several suitable dogs were also purchased from owners or dog pounds. Tests had shown that females were more sedate and less inclined to become fractious under pressure than males and so were selected for this reason. As well, the specially designed space suit they would wear was fitted with a device to collect urine and faecal matter, and was only manufactured to suit females. Each animal's height, length and weight were carefully recorded, and each was given a cute identifying nickname.[2]

Dr Nikolai Parin from the Soviet Academy of Medical Sciences once explained the reason behind the selection of canines as subjects for space travel. 'The Russian dog has long been a friend of science,' he stated. 'We have collected much information on our four-footed friends. Their blood circulation and respiration are close to man's. And they are patient and durable under long experiments.'[3] The selected dogs underwent preliminary testing and were divided into three categories according to their nature. The first group comprised even-tempered animals; the second, those who were restless and excitable; and finally there were the sluggish or slow dogs. The first school of even-tempered canines were destined for training specifically for long space flights and included such dogs as Strelka (Little Arrow), Belka (Squirrel), Lisichka (Little Fox), Chernushka (Blackie), Zhemchuzhnaya (Little Pearl) and Laika (Barker).

The dogs' breathing, heartbeats and temperatures were checked daily. They were X-rayed and taken on high-altitude aircraft tests to become familiar with the sensations of flight. Each was trained to wear a lightweight space suit, to be strapped into a confining canister and to eat off special trays. Through other tests, the dogs grew accustomed to the effects of vibration. They were then strapped into centrifuge

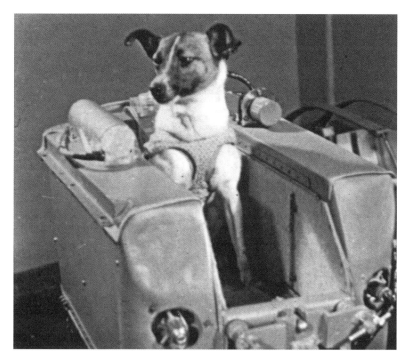

Laika looks over the capsule that will one day carry her into space.

chambers and whirled around at high speed to gauge their reaction to the forces of strong acceleration. Meanwhile, some of the other dogs were being trained for test ballistic space shots, touching on space, together with a menagerie of other small animals.

The first programme of six experimental R-1 rocket launchings carrying dogs took place between 1951 and 1952. Nine of the selected canines took part in these short, high-altitude geophysical flights, with three making two flights. The animals were checked post-flight and reported to be in good health. Nevertheless, Soviet scientists would not release details of these experimental rocket flights for several years; back then (and even to an extent now), most information was sketchy. Among the dogs known to have flown on these tests were Albina (Whitey), Dymka (Smoky), Modnista (Fashionable), Kozyavka (Gnat), Malyshka (Little One) and Tsyganka (Gypsy Girl).

It was then decided to advance the programme to a second phase, and eight more canines were recruited into the ranks. This second series of high-altitude rocket flights was conducted between 1955 and 1956, using more advanced R-1D and R-1Ye variants of the V-2. Twelve dogs took part, and once again some flew twice.

By now, special pressure suits had been developed for the animals, which included large spherical helmets. Their two-section capsule, specially built to hold two dogs, contained a movie camera together with a bright lamp and mirrors, so that the animals could be filmed from several angles. A layer of insulating material was applied to the walls of the split capsule to keep temperatures at a comfortable level for the occupants. The dogs had been trained to stand calmly while strapped securely into their sealed container for up to four hours.

Rocket flight veterans Albina and Tsyganka were chosen for the first in this new series of geophysical flights. Early on the morning of 26 March 1955, three hours before the planned lift-off, they were driven to the launch site, where they were inserted into their space suits and capsule. Following a successful launch, the rocket reached an altitude of 30 km (98,000 ft) before the fuel was exhausted, but the momentum caused it to streak upwards for several more seconds. When it reached a peak altitude of 87 km (285,000 ft), the right-hand section of the capsule, containing Albina, was jettisoned outwards. Soon after, Tsyganka's section was also ejected from the now descending nose cone. Three seconds after Albina was ejected, her parachute system deployed and billowed. She drifted slowly back to earth, breathing oxygen through her suit. Tsyganka also free-fell towards earth until her parachute opened as planned, 4 km (13,100 ft) above the ground. On landing, Tsyganka was nearly 20 km (12 mi.) from the launch site. Albina touched down a further 60 km (37 mi.) away.[4]

Both dogs not only survived their dramatic voyage but provided valuable data to scientists who had kept track of the animals' condition throughout the flight. Soviet physicians were quite amazed that they had remained so calm and had not attempted to escape from their straps. Recovered film showed that the two dogs had continually nodded their heads during the short period of weightlessness, while their blood pressure, heart and breathing rates had climbed a little at first before settling down to normal.

Other flights followed, and this method of ejecting the animals from their descending capsules proved highly successful – so much so that it was later adapted for use in safely returning the first cosmonauts to the ground. Given the rudimentary nature of this test flight programme it is difficult to believe that all of the dogs survived, but according to spokesperson Professor Alexei Pokrovsky, then director of Moscow's Institute of Aviation Medicine, 'none of the rocket flights during either the first or second stage of the work resulted in the death of animals.'[5]

A final series of seven high-altitude experimental flights took place between May 1957 and September 1960 atop the more powerful R-2A rockets, reaching altitudes of 200–212 km (656,000–696,000 ft). A total of fourteen dogs were launched, including Otvazhnaya (Brave One), who lived up to her name by making what is believed to be a total of five flights.

Belyanka (Snowy) and Pestraya (Piebald) soared to a record height on 27 August 1958 before parachuting back to earth. On 2 July the following year, another two dogs, Snezhinka (Snowflake) and Otvazhnaya, together with a rabbit named Marfusha (an affectionate form of the name Martha), were safely recovered after a suborbital flight recorded as travelling 'to a great height'.

Flight veteran Otvazhnaya and Malek (Tiny), along with a rabbit named Zvezdochka (Little Star), then made a successful high-altitude rocket flight on 15 June 1960. Just nine days later, Otvazhnaya was sent aloft for the fourth time, accompanied by another dog named Zhemchuzhnaya (Little Pearl).[6]

It had now become obvious to Soviet scientists through the flights and successful parachute recoveries of these 'animalnauts' that a human could survive similar expeditions into space, and they began to prepare for sending them on orbital test flights.

Meanwhile, on 4 October 1957, even as the final series of ballistic trajectory tests carrying dogs were taking place, the history of the world would change forever. That evening, Radio Moscow made the electrifying announcement that Iskustvennyi Sputnik Zemli ('Artificial Space Traveller around the Earth') had been launched and successfully inserted into orbit. Lift-off had occurred at 10:28:04 p.m. Moscow time. The Sputnik satellite separated as planned from the carrier rocket after orbital insertion, and a mechanism had been actuated which released its antenna. The world could now tune in to the satellite's primitive but historic beep-beep transmission and watch in awe – and not a little admiration – as it passed overhead.

Historically speaking, Sputnik was an appropriate name for the spacecraft. For millions of years, our Earth had been accompanied on its celestial travel by the Moon, nature's own satellite, but on that momentous day, Earth suddenly had two satellites. To that time, no man-made object had ever travelled faster than about 11,000 km/h

(6,835 mph), yet here was an impudent 58-centimetre (22.8 in.) aluminium sphere circling the globe at nearly three times that speed. The launch would occur during a global cooperative science initiative known as the International Geophysical Year (IGY), which lasted from 1 July 1957 to 31 December 1958.

Not only had a triumphant Soviet Union become the first nation to launch an artificial satellite into Earth orbit, but the success story that was Sputnik heralded the true dawning of the Space Age. Nevertheless, the stunning news produced a diverse crop of reactions. An effusive Sir Bernard Lovell, who was director of Jodrell Bank Observatory in Cheshire, called it 'absolutely stupendous . . . the biggest thing in scientific history . . . [and] the highest scientific achievement of the human intellect'.[7] Far less graciously, the U.S. Chief of Naval Research, Rawson Bennett II, called it 'a hunk of iron almost anyone could have launched'.[8] Manchester's *Guardian* newspaper was more circumspect:

> Somewhere along the line the honourable intentions of IGY space science have become submerged by national prestige, international conflict, and military security secrecy. Five years ago there was a great hope for the benefits which space flight could bring. Technical benefits are indeed beginning to appear, but cold-war attitudes are now proving a menace more dangerous than any radiation belt.[9]

President Eisenhower, never a great proponent of space science, merely humbugged the news. 'One small ball in the air', he grouched, adding that it 'does not raise my apprehensions, not one iota'.[10] He would soon change his mind, as there was genuine concern and even fear across the United States that Russia, which had a nuclear capability, might launch powerful rockets carrying warheads into orbit, which could be dropped on any target around the world.

If the sensational news that the Sputnik satellite was orbiting Earth proved disturbing to the people and scientists of America, then bulletins from Moscow just a month later, on the morning of 4 November, would prove equally devastating. A far heavier satellite named Sputnik 2 had been launched into orbit, and this time there was a passenger on board, the first living creature ever to make a true space flight: a dog named Laika.

News editors were sent into a frenzy as they tried to obtain details of the little space traveller. Some early newspapers humorously referred to the dog as 'Muttnik' until her name was released by the TASS news agency over Radio Moscow (although it later emerged she had originally been named Kudryavka, meaning Little Curly, by her trainers). Laika was identified as a female, 6-kilogram (13 lb) short-haired part-Samoyed terrier. Ten dogs had been considered for the flight, but in the end Laika was chosen for her gentle disposition, with Albina to act as her backup, and Myshka as the support dog during a number of flight-related tests. During Laika's orbital flight there was much speculation in the Western press as to when – or if – she would be brought back to Earth. Soviet scientists maintained a strict silence over the eventual fate of the dog, but the sad truth is that little Laika was sent on her ill-fated journey into space with no means of returning home.

In spite of their growing experience with ballistic rocket flights, the Soviets simply did not have the capability at that time to return an animal from orbit. Therefore, in Sergei Korolev's planning for the flight of Sputnik 2, no allowance was made for the chosen dog's recovery. Her sealed capsule was designed to sustain and keep her alive for a little over a week. Sufficient food and water would be supplied from a dispenser long enough for data to be gathered on her condition and behaviour in orbit. To provide breathable air, scientists installed a tank of oxygen and a regenerator to clear the capsule of excess (potentially fatal) carbon dioxide and moisture, which was then reconverted to oxygen.

On the day of her historic mission, Laika was carefully groomed by attendants, who sponged and combed her coat. Electrodes were attached on and beneath her skin to monitor her respiration and heart-beat – data that would be transmitted back to Earth. A rubber bag was attached to her rear end to collect body wastes so they would not be floating free in the weightless environment of the capsule. Through all of this, Laika remained placid and quiet. It was a procedure she had endured many times before.

After a special harness had been fitted, Laika was placed into a narrow, padded canister and small chains were secured to keep her in place. These chains did not allow her to move very far, but she was able to lie down, sit or stand.

Laika's capsule was then fitted into the 4-metre-high nose section of the R-7 rocket and the countdown began. Lift-off went as planned at 5:30 a.m. on 4 November 1957, and soon Sputnik 2 and Laika were soaring into the dawn sky. After reaching orbit, the nose cone containing Laika separated as planned from the carrier rocket, although a problem seems to have resulted during the violent explosive separation. According to noted Soviet space researcher Anatoly Zak, 'telemetry later revealed Laika's heart was beating 260 cycles per minute, or three times higher than normal during the ride to orbit. Frequency of her breath also rose four to five times above usual. Overall, however, the dog survived the launch unscathed.'[11]

Although Sputnik 2 had achieved orbit, telemetry relayed to the ground showed an unexpected and critical rise in the cabin temperature. Some of the thermal insulation within Laika's capsule had apparently been torn loose during the separation process. This prevented the thermal control system from functioning correctly, raising the cabin temperature to 40°C (104°F). Laika's vital signs indicated that she was becoming anxious but that she had otherwise suffered no ill effects during the launch and insertion into orbit. Although she began to feed from the automatic dispenser, she grew increasingly restless as the temperature climbed above 40°C in her cramped, confining capsule. According to scientists monitoring her vital signs back on Earth, she began barking and moving about in an agitated way.

Despite these problems, it was decided to announce the successful launch of the first living creature into orbit. Laika became an instant celebrity around the world, with many newspapers lauding her as 'the most famous dog in history'. Further announcements from Moscow were brief, but all stated that Laika was calm, behaving normally, and that her condition was 'satisfactory'. Understandably, her presence on the space flight, and persistent rumours that she would ultimately perish in space, angered dog lovers around the world.

In several countries, people protested about the use of a live animal in these experiments and wrote strongly worded letters to newspapers, condemning the Soviet Union's lack of compassion. In London, the National Canine Defence League urged dog lovers everywhere to observe a minute's silence each day on behalf of Laika. The Singapore Canine Welfare Association sent a telegram to Moscow demanding that 'a Russian hero' be sent next time 'instead of dumb and defenceless animals'.

Despite this, people around the world went outside each night to watch as Sputnik 2 raced across the sky, and prayers for the safety of its canine occupant followed the bright dot of light until it disappeared below the horizon. As the days passed, it became increasingly obvious – as speculated – that the Soviets had no plans to bring Laika back home, while impassive official statements declined to comment on the subject of her survival. After ten days in orbit, they reported that Laika's oxygen had run out, and she had peacefully slipped into final unconsciousness.

In truth, Laika quite probably died very early in her flight, and most likely in a great deal of distress due to the unbearably hot conditions inside her capsule. The sad fate of Laika was first revealed to the author during a chance meeting in Vienna in 1993 with Russian academician Oleg Gazenko, a doyen of early Soviet bioastronautics. During a Space Explorers' Congress held in that city, I explained to Dr Gazenko that I was writing about early Soviet dog flights and wanted to know whether Laika had perished when her air ran out. He revealed to me that Laika had probably died of heat prostration somewhere around four hours into her orbital flight and explained how this had occurred. These facts were later confirmed to me in a separate interview with pioneering cosmonaut Vitaly Sevastyanov.[12]

Sputnik 2 went on to make a total of 2,370 orbits before it plunged back into the atmosphere on the evening of 13 April 1958 and was incinerated in the fiery heat of re-entry. Following the flight of Sputnik 2, leading Soviet scientist Oleg Ivanovskiy (writing under the pseudonym Alexei Ivanov) stated: 'Laika's flight made it possible to speak more boldly and concretely of the possibility of cosmic journeys of men.'[13]

### Project Albert and the Astrochimps

Meanwhile, in the United States, Wernher von Braun and his team of Peenemünde scientists had settled in well with their families and were continuing their work in rocketry. Given access to a limited number of captured V-2 missiles with warheads removed and a huge stockpile of parts, their objective was to take the first steps in America's goal of human space exploration.

The rocket men knew that one of their eventual objectives was to launch a man into space, but they would first have to overcome some elementary physics. This meant increasing the velocity of their rockets from 5,000 km/h (3,100 mph) – the top speed ever reached by a V-2 missile – to somewhere around 29,000 km/h (18,020 mph), achieving sufficient velocity to overcome gravitational forces and accomplish orbital status.

The first launch of a v-2 rocket from American soil took place in 1946 at Holloman Air Force Base in New Mexico. At that time, the fastest any human had ever travelled in an aircraft was 975 km/h (606 mph), a record set on 17 November 1945 by RAF officer Hugh Wilson in a Gloster Meteor jet aircraft. Given the dynamic forces of launch involved in a ballistic rocket flight above the atmosphere, scientists were concerned that a person might not survive even a brief flight into space. Questions were raised as to how the human body might react to these unfamiliar forces and to the silent, malignant dangers of cosmic radiation lurking in the upper atmosphere.

These questions needed to be answered before the first human could safely penetrate outer space. Researchers agreed that the solution rested in the use of warm-blooded animals, whose physiology closely resembled that of human beings. Historically, animals had been used in high-altitude testing since 1783, when the Montgolfier brothers carried aloft a sheep, a duck and a rooster in one of their hot-air balloons to see if ground-dwelling animals could survive at high altitude. Their eight-minute flight reached around 457 m (1,500 ft) altitude, and everyone on board landed safely.

There was, however, still a great deal of preparatory work and research to be carried out in a host of laboratory and field tests, as well as aboard ever-more-advanced high-altitude balloons. Insects, spores, plant life and other specimens were launched on rocket flights, occupying cramped nose cones.

Telemetry on the first v-2 launches yielded vital information to scientists and von Braun's rocket team, but even with telemetry, certain biological payloads and items of equipment needed to be recovered intact. Development therefore began on creating reliable parachute systems to bring these nose cones and capsules safely back to Earth.

Unless some microbes managed to make their way unnoticed into the nose cone of an earlier rocket, the first living creatures launched into space from American soil were fruit flies. These were placed into the nose cone of a v-2 rocket that lifted off from White Sands Missile Range in New Mexico on 20 February 1947, reaching an altitude of 108 km (354,000 ft). Following a parachute landing, the insects were still alive.

Several v-2 rockets were allocated to the U.S. Air Force Cambridge Research Center for a programme called Project Blossom, which involved a series of test flights designed to explore the possibilities of ejecting a capsule and recovering it by means of a parachute. This effort was based at Holloman Air Force Base in Alamogordo, New Mexico, and the nearby precursor to Florida's Cape Canaveral – the U.S. Army's White Sands Proving Ground.

Following two successful launches in the Project Blossom series, the Air Force wanted to verify if humans could survive a flight into space. Of principal concern were the crushing g-forces associated with launch, the suspected disorientation that would be caused to a human passenger by weightlessness, and the dynamic effects of descent and landing. With this in mind, the Aero Medical Laboratory at Wright Air Development Center in Dayton, Ohio, was asked to supply 'simulated pilots' for upcoming Project Blossom launches. This project was to be headed by Colonel James P. Henry, director of the Air Force's Physiology of Rocket Flight research project, together with Captain David Simons as his project engineer.

The Blossom 3 launch had been scheduled for June 1949, giving Henry and Simons just two months to find a suitable candidate and design a life-supporting capsule. They quickly determined that rhesus macaques were best suited to the task. They were used in these early biological flights due to space limitations within primitive nose cone capsules, as they could be trained to perform simple repetitive tasks, and because physiologically they were very similar to human beings.

Providing an environment in which the animal would be able to travel to the thresholds of space and survive was a far more complex problem. The basic capsule requirements called for an oxygen system that would last from the time the capsule was sealed until its post-flight recovery; a chemical-based system to absorb the primate's carbon dioxide exhalations; and monitoring instruments to relay the animal's heartbeat and respiration back to the ground. All this had to be included in an aluminium capsule large enough to hold the primate, but small enough to fit into the odd-shaped space among instruments already crammed into the rocket's nose cone. Finally, a rudimentary space capsule was ready in time for the Blossom 3 launch. This pioneering series of primate flights was codenamed 'Albert', the name given to the rhesus monkey selected for the first test.

This flight, on 11 June 1948, ended in failure. Not only did the single parachute fail to inflate, causing the nose cone to slam back into the ground, but it was later revealed that Albert had probably suffocated before lift-off. Although the second primate, Albert II, survived his Blossom flight the following year, reaching an altitude of 56.3 km

Anaesthetized, the first Albert is inserted into his capsule ahead of the Blossom 3 launch.

(185,000 ft), the little primate perished through another parachute failure. At the eleven-second point after lift-off of the next Blossom flight three months later, a small explosion occurred in the tail of the v-2, followed fourteen seconds later by a much larger explosion, which blew the v-2 carrying Albert III to pieces. The last v-2 flight on 8 December 1949, with Albert IV aboard, proceeded well, but once again the parachute system failed, and the nose cone crashed into the ground at high speed.

With the last of the allocated v-2s used, attention then turned to the newly developed and lighter Aerobee sounding rockets. On 18 April 1951, Albert v was launched aboard an Aerobee rocket, achieving an altitude of 56.3 km (185,000 ft). Despite many months spent modifying the parachute system, it failed once again, and Albert v, together with a collection of mice, died on impact. It was a frustrating time for the research team, but a breakthrough was imminent with the development of a two-parachute system. The first drogue chute was 2.5 m (8.2 ft) in diameter and comprised of ribbons covered in metallic mesh to aid in radar tracking. It was designed to reduce the initial descent speed, following which a second, 4-metre (13 ft) chute would be deployed.

Aeromed Aerobee 2 was launched from the Holloman pad on 20 September 1951, reaching an altitude of 71 km (233,000 ft). The nose cone carrying Albert VI in his capsule along with eleven mice separated as scheduled and began its descent to Earth. The newly modified parachute system operated perfectly, and the nose cone floated down under a billowing canopy. Although it landed with a jarring thump, telemetry emanating from the desert indicated that Albert VI had survived the flight. He was eventually recovered, but the sun's blistering heat had proved too much for the little monkey; two hours after he was extracted from his capsule, he died of heat prostration. Despite

Preparations for the launch of rhesus monkey Albert I on the Blossom 3 V-2 rocket. Unfortunately the primate seems to have suffocated prior to lift-off on 11 June 1948.

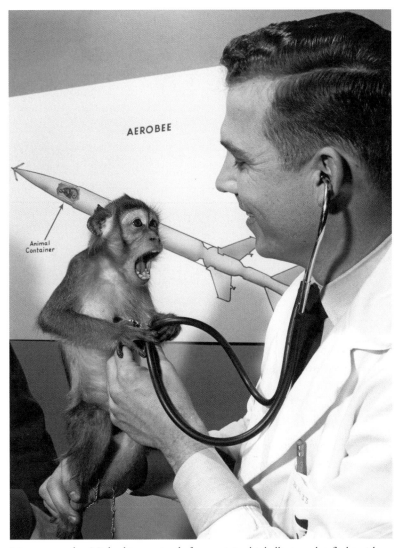

Macaque monkey Michael is examined after surviving his ballistic rocket flight with Patricia. A cutaway of the Aerobee rocket is in the background.

this major disappointment, there was encouragement in the fact that the parachute system had worked as planned.

The final flight in the series was launched on 21 May 1952. This time, two crab-eating macaques named Patricia and Michael were on board, together with two white mice, Mildred and Albert. Everything went well, although the Aerobee only reached a peak altitude of 63 km (207,000 ft). The parachute system worked perfectly, and all of the animals were recovered alive. This flight would mark the last American

## V-2 Rocket Biological Flights

| Launch Vehicle | Launch Date | Launch Site | Biological Payload | Approx. Altitude / Test Summary |
|---|---|---|---|---|
| V-2 No. 7 | 9 July 1946 | WSPG | Specially developed strains of seeds | 134 km / Samples not recovered |
| V-2 No. 8 | 19 July 1946 | WSPG | Specially developed strains of seeds | 6.4 km / Samples not recovered |
| V-2 No. 9 | 30 July 1946 | WSPG | Ordinary corn seeds | 167 km / Seeds recovered |
| V-2 No. 12 | 10 October 1946 | WSPG | Rye seeds | 180 km / Seeds recovered |
| V-2 No. 17 | 17 December 1946 | WSPG | Fungus spores | 188 km / Spores not recovered |
| V-2 No. 20 | 20 February 1947 | WSPG | Fruit flies | 109 km / Flies recovered alive |
| V-2 No. 21 | 7 March 1947 | WSPG | Rye seeds, corn seeds, fruit flies | 163 km / Biological result not recorded |
| V-2 No. 29 | 10 July 1947 | WSPG | Rye seeds, corn seeds, fruit flies | 16 km / Biological result not recorded |
| V-2 No. 37 Blossom 3 | 11 June 1948 | WSPG | Monkey (Albert) | 63 km / Parachute failure; animal died |
| V-2 No. 44 | 18 November 1948 | WSPG | Cotton seeds | 145 km / Seeds recovered |
| V-2 No. 50 | 11 April 1949 | WSPG | Not known | 87 km / Biological results not recorded |
| V-2 No. 47 Blossom 4B | 14 June 1949 | WSPG | Monkey (Albert II) | 134 km / Parachute failure; animal died |
| V-2 No. 32 Blossom 4C | 16 September 1949 | WSPG | Monkey (Albert III) | 4.8 km / Rocket exploded; animal died |
| V-2 No. 31 | 8 December 1949 | WSPG | Monkey (Albert IV) | 132 km / Animal died on impact |
| V-2 No. 51 | 31 August 1950 | WSPG | Mouse | 135 km / Animal died on impact |

## Aerobee Biological Rocket Flights

| Aerobee No. | Launch Date | Launch Site | Biological Payload | Approx. Altitude / Test Summary |
|---|---|---|---|---|
| USAF-12 | 18 April 1951 | WSPG | Monkey (Albert V), several mice | 61 km / Parachute failure; animals died |
| USAF-19 | 20 September 1951 | WSPG | Monkey (Albert VI), 11 mice | 71 km / All animals recovered; monkey died two hours after impact |
| USAF-20 | 21 May 1952 | WSPG | Two monkeys (Patricia and Michael), two mice | 63 km / All animals safely recovered |

research flight using primates for six years, and the knowledge drawn from these pioneering flights was crucial in formulating future plans.[14]

After becoming a U.S. citizen, Wernher von Braun was appointed civilian head and technical director of the Army Ballistic Missile Agency (ABMA) in Huntsville, Alabama, then under the command of Major General John B. Medaris. The ABMA, established at the Redstone Arsenal on 1 February 1956, was charged with developing the U.S. Army's first large ballistic missile, the Jupiter. Von Braun's team of German rocket scientists were playing a vital role in the missile testing and development programme.

As von Braun began to assert his rapidly growing influence, he proposed the development of a national space agency, with an annual budget of $1.5 billion, that would undertake programmes designed to produce 'a man orbiting the earth on a returnable basis' within five years, and a manned space station in ten years.[15] Medaris was mostly in agreement with von Braun, stating before a Defense Department subcommittee on 14 December 1957 that there was an urgent need to accelerate the Atlas intercontinental ballistic missile programme, testifying that, 'Unless we develop an engine with a million-pound thrust by 1961, we will not be in space; we will be out of the race . . . The priority today should be on the attainment of a space capability at the earliest possible date.' He then added, 'Right now we need a ten- to twelve-year program that has as its ultimate goal the manned domination of space.'[16]

Garrison Norton, then Assistant Secretary of the Navy (AIR) overseeing naval aeronautics, told the subcommittee two days later that he outright rejected the 'Buck Rogers' notion of developing manned spacecraft and testified that the service's number-one priority should be to develop a missile 'to get a warhead from here to its target with accuracy. Nothing should dilute that effort', he declared. 'I do not belittle the importance of satellites and space matters one bit, but what I want to emphasize is, let's get first things done first.' He also charged that the Navy's missile research programme had been seriously hampered by negative publicity and by a 'budgetary straitjacket'.[17]

On 6 December 1958, right across the United States, people tuned in to their radio sets, expecting to hear a delayed bulletin from Cape Canaveral. That day, America stood ready to become a true space nation by sending a compact U.S. Navy satellite named Vanguard into orbit – big news after the Sputnik debacle. Despite this, there would be no direct radio broadcast of the launch, nor any live TV transmission; in fact, no images of the launch would be seen until a filmed report due to be screened on that night's news. Not only was the u.s. military particularly secretive and sensitive about allowing its launches to be shown live across the nation, but at the time there were simply no direct-transmission lines available. As the countdown reached zero, the carrier missile roared into lusty life, rose a little off the launch pad, shuddered, and was suddenly engulfed in a massive orange fireball. The much-publicized effort to place Vanguard into orbit had ended in a morale-shattering anticlimax. The team's propulsion engineer, Kurt Stehling, would later write that, as the countdown passed zero:

The Vanguard launch ended in a massive launch pad explosion.

It seemed as if all the gates of Hell had opened up. Brilliant stiletto flames shot out from the side of the rocket near the engine. The vehicle agonizingly hesitated a moment, quivered again, and in front of our unbelieving, shocked eyes, began to topple. It also sank like a great flaming sword into its scabbard down into the blast tube. It toppled slowly, breaking apart, hitting part of the test stand and ground with a tremendous roar that could be felt and heard even behind the two-foot concrete walls of the blockhouse.[18]

Following this demoralizing failure, Washington hurriedly authorized the ABMA in Huntsville to launch an artificial satellite into orbit using its reliable Jupiter-C research rocket. As the agency's director, von Braun was instructed to manage this, and he confidently predicted he would complete the task within ninety days.

The U.S. Army, von Braun and his rocket team were successful: on 31 January 1958, a Jupiter-C rocket left the launch pad and soared without incident into the heavens. Minutes later, the United States finally had its own satellite, Explorer 1, in orbit.

### A Civilian Space Agency

By October 1958, the worldwide programme of cooperative research known as the International Geophysical Year was coming to an end. Begun in July 1957, the IGY had involved a systematic study of Earth and its planetary environment. During this time the Soviet Union successfully orbited three Sputnik satellites, while the United States had responded by orbiting three of its own, albeit much smaller, satellites. Ostensibly, it seemed that the Russians were well ahead in space technology, both in rocketry and in their ability to launch large, complex spacecraft into orbit.

One of the major problems on the American side of things had been the matter of inter-service rivalry between the U.S. Navy, with its flawed Vanguard programme, and the U.S. Army, then preparing to launch the first of its Discoverer satellites into polar orbit atop a Thor–Agena combination launch vehicle, comprised of a Thor first-stage rocket and an Agena second stage to give the final boost into orbit. In January 1958 the Department of the Army had proposed a joint effort by the three services to send a man into space under the working project title 'Man Very High', but in April the Air Force indicated that it no longer wanted to participate, while the Navy was becoming increasingly lukewarm on such a venture. Undeterred, the Army pressed on with its programme, now redesignated Project Adam, with plans to send a human candidate on a ballistic flight to an altitude of around 250 km (155 mi.) in a recoverable capsule mounted atop a Redstone rocket. By July, however, those plans had been consigned to the scrap bin as unworkable. It was against the background of this intense inter-service rivalry that a civilian space agency, the National Aeronautics and Space Administration (NASA), was created.

On 8 August that same year, President Eisenhower appointed T. Keith Glennan – then president of the Case Institute of Technology in Ohio – as the first administrator of the new non-military space agency, with an initial budget of $330 million. Most of the personnel and resources for NASA were drawn from the agency it would replace, the National Advisory Committee for Aeronautics (NACA). At the same time, all existing military space projects, including projects Explorer and Vanguard, were placed directly under the administration of NASA. Glennan's deputy was named as Hugh L. Dryden, the last director of NACA.

Within a week of his own secondment from NACA, aerospace engineer Robert R. Gilruth had been appointed to assemble and head a select committee known as the Space Task Group (STG), headquartered at the Air Force's Langley Field, Virginia. They were charged with investigating the possibility of putting together a civilian programme to orbit a manned spacecraft, study the effects of weightlessness on humans, investigate a pilot's ability to function in space, and to safely recover both pilot and spacecraft – ideally before the Soviet Union. The programme received official approval on 7 October. Eisenhower had decided that the name of the project, originally designated Project Astronaut by the STG, conveyed far too much emphasis on the pilot and not the programme. Project Mercury was the alternative name suggested by committee member Abe Silverstein, in honour of the mythological Roman messenger of the gods. The name was formally adopted on 17 December 1958, as was the term 'astronaut', following Gilruth's suggestion. As mentioned in *Origins of NASA Names*, 'The term followed the semantic tradition begun with "Argonauts," the legendary Greeks who travelled far and wide in search of the Golden Fleece, and continued with "aeronauts" – pioneers of balloon flight.'[19]

NASA's deputy administrator Hugh Dryden (left), President Dwight D. Eisenhower and the newly formed space agency's first administrator T. Keith Glennan formalize the creation of NASA on 1 October 1958.

As far as suitable launch vehicles were concerned, NASA chose to modify existing military ballistic missiles to deliver its payloads to space. The Army's Redstone missile could be adapted for use on manned suborbital flights, while a larger and far more powerful rocket, the Atlas D, was under development for the U.S. Air Force. The Atlas could be modified to enable manned orbital missions.

Unlike the Soviet Union, NASA decided on an open policy, without the complete secrecy of the kind then shrouding the Soviet space programme. This openness would create problems, however, with Russian cosmonauts later flying missions deliberately planned to pre-empt and better every early American manned mission, while creating enormous propaganda value for the Soviet space effort. NASA's open policy would nevertheless work to enthuse and inform Americans, who took pride in the nation's space flight programme and the men chosen to fly them. By way of contrast, early Soviet missions would not be announced until after they had been successfully launched, while the names of the cosmonauts involved were a closely guarded state secret before they flew. It would be many years before this policy was finally relaxed.

On 13 December 1958, as part of a NASA biological programme, a small South American squirrel monkey called Gordo was launched from the Atlantic Missile Range at Cape Canaveral, Florida. The monkey was snugly encased in a small capsule within the nose cone of a U.S. Army Jupiter AM-13 rocket, which had originally been developed to carry intermediate-range missiles. Telemetry relayed to the ground indicated that Gordo was reacting well during the flight, the only minor side effect being a slight slowing of his heart rate. During his fifteen-minute suborbital flight, Gordo was weightless for 8.3 minutes. Unfortunately, a technical malfunction after re-entry prevented the parachute from opening, and the nose cone apparently sank after splashing down in the North Atlantic. Despite a desperate six-hour sea search by the Army, the nose cone could not be located. Otherwise, the mission was deemed a success.[20]

Determined to press on with their biological efforts, NASA chose two monkeys for its next mission in 1959. The first was a female rhesus macaque named Able, selected out of a group of 24 candidates recruited from a zoo in Independence, Kansas. The second of the flight primates was a much smaller female, a member of another large candidate group purchased from a pet store in Miami. She was a two-year-old South American-born squirrel monkey weighing less than half a kilogram (1 lb). In keeping with the naming of the larger primate as Able, the first letter in the military's phonetic alphabet, the tiny monkey was in turn named Miss Baker.

During their training, each of the monkeys was dressed in a specially manufactured suit fitted with sensors to track their pulse rate, body temperature and movement. They would be strapped into separate capsules within the nose cone during their flight. While Miss Baker would be completely encased in a tiny capsule the size of a thermos flask, Able had enough room to perform a simple, repetitive task during the flight. She was trained to press a button with her finger whenever a red light began flashing, allowing scientists to test her coordination and focus during the flight and while experiencing a nine-minute period of weightlessness. Despite the presence of the two primates and other biological experiments, they were of secondary importance to the primary purpose of the flight, which was to test the nose cone's ability to protect its payloads. The nose cone would be subjected to temperatures of around 2,800°C (5,000°F) as it returned through the ferocious heat of re-entry.

On 28 May 1959, an Army-built Jupiter AM-18 blasted into the sky from the Atlantic Missile Range with the two primates secured in the nose cone. This NASA mission was also planned to test the effects of cosmic radiation, increased gravity and weightlessness on sea urchin eggs, human blood cells, yeast and onion-skin cells, corn and mustard seeds, mould spores and fruit fly larvae. In a flight lasting nearly seventeen minutes, the nose cone travelled some 2,735 km (1,700 mi.) from Cape Canaveral and reached an altitude of approximately 579 km (360 mi.). Following a fiery re-entry, the nose cone plunged into the ocean about 400 km (249 mi.) southeast of San Juan, Puerto Rico. Initially, the ocean recovery team feared that the nose cone had sunk, much like the earlier flight of Gordo, but then it was spotted bobbing in the water and quickly recovered by Navy frogmen a little over an hour and a half after lift-off. Shortly thereafter, a message came through to the Cape

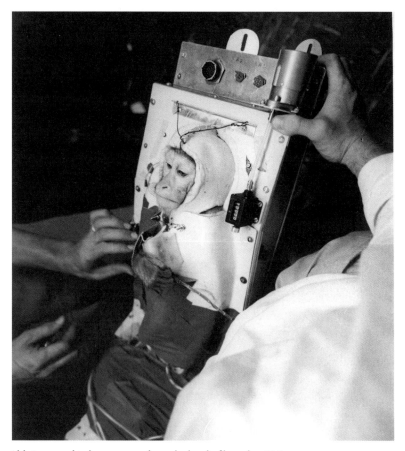

Able is secured in her contoured couch ahead of launch, 28 May 1959.

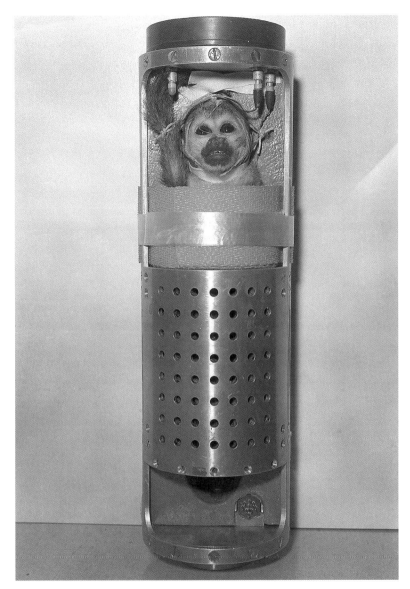

Tiny squirrel monkey Miss Baker secured inside her capsule, prior to launching with her female companion Able on a Jupiter missile.

Canaveral control room: 'Able Baker perfect. No injuries or other difficulties.' A wealth of useful biological data had been gathered during the animals' fifteen-minute flight, including body temperature, muscle reaction and respiration rate. It was yet another milestone in proving that there were no substantial impediments to sending humans into the space environment.

43

Two days after their flight, NASA administrator T. Keith Glennan displayed Able and Baker at a Washington press conference, stating that the two monkeys had withstood forces up to 38 times that of gravity during their flight, while their heartbeats and respiration had changed only moderately. Photos of Able and Baker would appear on the cover and within the pages of the subsequent issue of *Life* magazine.

Although the two monkeys survived their space flight in excellent condition, Able died on 1 June, just four days after her historic journey. It followed a reaction to the anaesthetic given during surgery by Army surgeons to remove an infected electrode implanted in her abdomen pre-flight. Miss Baker, who became an instant celebrity after her one and only space flight, survived until November 1984, dying of kidney failure in Huntsville, Alabama, aged 27.

The race between the Soviet Union and the United States to place the first person into space was gaining momentum as animal space shots continued to return vital data needed before strapping a man into a spacecraft. On 15 May 1960, the prototype of a new Soviet spacecraft designated Vostok – officially Korabl-Sputnik 1, translating as Spaceship-Satellite 1 – was launched from the Baikonur Cosmodrome in southern Kazakhstan, carrying a mannequin cosmonaut. The spacecraft had no heat shield, as the mission was purposely intended to end with its destruction on re-entry. The Vostok craft circled the planet for four days as verification of the spacecraft's systems was conducted from the

### Jupiter Biological Rocket Flights

| Launch Vehicle | Launch Date | Launch Site | Biological Payload | Approx. Altitude and Distance / Result Summary |
|---|---|---|---|---|
| Jupiter IRBM AM-13/ Bio-flight No. 1 | 13 December 1958 | Cape Canaveral | Monkey (Old Reliable, aka Gordo) and neurospora | 467 km altitude, 2,092 km distance / Nose cone lost at sea |
| Jupiter IRBM AM-18/ Bio-flight No. 2 (2A and 2B) | 28 May 1959 | Cape Canaveral | Two monkeys (Able and Baker), neurospora, seeds, pupae, sea urchin eggs | 483 km altitude, 2414 km distance / Nose cone and both animals safely recovered |
| Jupiter IRBM AM-23 | 16 September 1959 | Cape Canaveral | Two frogs, twelve pregnant mice, seeds, pupae, urchin eggs | Missile destroyed by range safety officer after flawed launch |

ground. It would remain in orbit for the next five years before its orbit finally decayed and it was incinerated during the fiery re-entry. Despite some technical problems, the mission was deemed a success, and a follow-up 24-hour flight was scheduled that would carry a small menagerie of animals, including two dogs. If everything went well, they would become the first living creatures recovered from orbit.

A pair of Russian dogs named Lisichka (Little Fox) and Chaika (Seagull) were subsequently launched in a second prototype Vostok spacecraft from Baikonur on 28 July 1960. Misfortune would strike early when a strap-on booster broke away from the R-7 rocket during launch and it began breaking up nineteen seconds after lift-off. Ground-based flight controllers quickly initiated a command to jettison the shroud covering the ascending payload, following which the Vostok's descent module, containing the two dogs, was explosively separated from the booster. Parachutes were deployed, but at far too low an altitude. Both canines died when their spacecraft slammed into the ground at high speed.

Following an investigation into the failure, a decision was reached to develop an ejector seat that would allow any future cosmonauts to make a rapid escape from their Vostok vehicle in the event of a similar launch failure, as the parachutes provided for the spacecraft would not fully deploy and billow until about 40 seconds after lift-off. The following month, on 19 August, another Vostok spacecraft, Korabl-Sputnik 2 (incorrectly called Sputnik 5 in the West), was launched into orbit from Baikonur atop a Vostok-L carrier rocket. On board were two dogs named Belka (Whitey) and Strelka (Little Arrow), as well as a grey rabbit called Marfusha, 42 mice, two rats, several fruit flies and some plants and fungi. A television camera was also installed to record the reactions of the two dogs during their flight.

Following their day-long orbital mission, the spacefaring animals were recovered safely. Most notably, Belka and Strelka had become the first dogs to enter space and return alive after their orbital mission, giving the Soviets increased confidence in their bid to send a cosmonaut on a similar journey. The two small dogs and their achievement were feted around the world. A year later, Strelka had a litter of puppies, one of which, Pushinka (Fluffy), was presented to U.S. First Lady Jacqueline Kennedy by Soviet premier Nikita Khrushchev as a gift for her daughter, Caroline, who was delighted with the little puppy. Pushinka would later have puppies of her own – Blackie, Butterfly, White Tip and Streaker

Belka (left) and Strelka, the first living animals returned to Earth after an orbital space flight.

– which President Kennedy jokingly referred to as 'pupniks'. The four puppies were later given away.

Korabl-Sputnik 3 was successfully launched on 1 December 1960, but unlike the triumphant preceding flight of Belka and Strelka, this one ended in disaster. In addition to the dogs Pchelka (Little Bee) and Mushka (Little Fly), the other specimens aboard the spherical Vostok craft were guinea pigs and rats, plants and other biological experiments. Different breeds of mice and fruit flies had also been loaded in order to study the effects of cosmic radiation. As before, television cameras were fitted to record the behaviour of the dogs. The one-day flight of Korabl-Sputnik 3 went well until the re-entry phase, when the retrorocket designed to slow the craft malfunctioned. It was decided to fly an additional orbit before the spacecraft began its descent through the atmosphere, this time on an unplanned trajectory. Official Soviet media sources would announce that Korabl-Sputnik 3 and its occupants had been lost on re-entry. However, it was later revealed that the spacecraft had been deliberately destroyed when a self-destruct mechanism, fitted in case the spacecraft landed outside the territory of the Soviet Union and so could be retrieved and studied by agents of foreign powers, was activated from the ground.

Three weeks later, on 22 December, the fifth test flight of a Vostok capsule began badly. On this occasion the r-7 rocket had been modified with the addition of a third stage boasting a higher thrust. The two dogs on board were named as Kometa (Comet) and Shutka (Joke), but they were in for a short and troubled flight. Seven minutes into the ascent, the third-stage engine cut out prematurely. Once this happened, the emergency escape system was activated. The service module did not separate from the spherical re-entry capsule as designed but remained connected by a wire bundle that only burned away during re-entry. The capsule landed in a remote area near the Lower Tunguska River in Siberia. Even with the aid of a recovery beacon it still took several days for recovery teams to locate the spacecraft, in freezing temperatures. Remarkably, the two dogs were found to be still alive, and were retrieved. The spacecraft was returned to Moscow a few days later. Because it failed to achieve orbit, the flight was never allocated a Korabl-Sputnik designation.

On 9 March 1961 a lone female dog, Chernushka (Blackie), was launched aboard Korabl-Sputnik 4 in a small, pressurized sphere, together with eighty mice, numerous guinea pigs and a number of biological specimens. Of more interest to the research team was the fact that this flight carried a 'passenger' – a space-suited wooden mannequin that Soviet officials had given the tongue-in-cheek name of Ivan Ivanovich. The mannequin was designed to replicate the size, weight and mass of a real person and was strapped in and clothed in the same orange space suit that would be worn by the first cosmonaut. Housed inside Ivan's chest cavity, abdomen and hip area was a veritable Noah's Ark of other life forms, including additional mice, guinea pigs and microorganisms. As planned, the spacecraft made a single orbit of the planet, following which the descent module made a successful re-entry through the atmosphere. Chernushka was recovered in good condition and spirits. 'Ivan' was also in good shape after being automatically ejected from the capsule and landing by parachute. The entire mission was rated a complete success and was cause for greatly increased optimism and excitement as the Soviets now prepared for their first manned flight.[21]

The flight profile of the Korabl-Sputnik 4 mission had been the same as the one then being developed for the first cosmonaut, following a final test flight of the Vostok spacecraft and its systems. This crucial mission began on 25 March 1961 from the Baikonur Cosmodrome with the successful launch of the Korabl-Sputnik 5 spacecraft. Once again, a single dog was on board, accompanied by 'Ivan Ivanovich' on his second mission into orbit and another menagerie of small animals and biological samples. Testing the communication link from the spacecraft to ground control was an important feature of this flight, so the capsule had been fitted with a recorder. During the one-orbit flight it would play a recording of the Piatnitsky Choir, followed by a voice reciting a recipe for cabbage soup. Not only would it test the communications system, but if Western ears happened to tune in to the recordings it would certainly cause a great deal of confusion. The little dog selected for this vital mission was named Zvezdochka.

There would be six important witnesses to the launch of Korabl-Sputnik 5 at Baikonur – the six cosmonauts selected as the best to fly the first series of manned Vostok missions. Known as the Vanguard Six, they were Yuri Gagarin, Gherman Titov, Andrian Nikolayev, Pavel Popovich, Valery Bykovsky and Grigori Nelyubov. Earlier, they had met the small dog who would fly aboard the spherical capsule. One of the cosmonauts, Gagarin, is said to have asked the handler for the canine's name, but the man did not know, guessing that it was either Dymka (Smoky) or Tuchka (Cloudy). Gagarin was aghast and said the dog should have a name better befitting the occasion, suggesting Zvezdochka, which translates to 'Little Star'.

The Vostok craft successfully slipped into low Earth orbit and completed a single orbit before re-entering the atmosphere over the Soviet Union. As the capsule hurtled towards the ground, the mannequin's couch was ejected as scheduled and, duplicating the previous mission, 'Ivan' touched down beneath his parachute. The spacecraft, under its own parachute, landed nearby, northeast of Izhevsk in the western Ural Mountains. Both spacecraft and mannequin had landed in the middle of a snowstorm, which meant that it would be some 24 hours before the

## Precursory R-7 Canine Orbital Flights

| Launch Date | Dog(s) | Result |
|---|---|---|
| 28 July 1960 | Chaika and Lisichka | Rocket malfunction after launch; both dogs perished |
| 19 August 1960 | Belka and Strelka | Orbited, returned safely |
| 1 December 1960 | Pchelka and Mushka | Died in re-entry explosion |
| 22 December 1960 | Kometa and Shutka | Orbited, returned safely |
| 9 March 1961 | Chernushka | Orbited, returned safely |
| 23 March 1961 | Zvezdochka | Orbited, returned safely |

recovery team could reach the site and prepare them for return to Moscow. Zvezdochka was in perfect health after her orbital adventure.

The Soviet space chiefs now felt that they could safely commit to launching a cosmonaut on a similar, single-orbit flight around the Earth. Just three weeks later, that determination paid off, with the first human space explorer finally taking over from the pioneering flights – and often sacrifices – of a steady stream of Russian canines.

The Mercury astronauts pose with an Atlas rocket model. Front row (l to r) Gus Grissom, Scott Carpenter, Deke Slayton, Gordon Cooper. Back: Alan Shepard, Wally Schirra and John Glenn.

# 2

# FIRST INTO OUTER SPACE

I t was 1958, and for America the time had finally come to roll back the curtain on the unknown. As a jubilant Soviet Union continued to celebrate and gloat over the achievement of the world's first artificial satellite, followed by the launch of a small dog named Laika into Earth orbit, the fledgling space agency NASA was also directing its attention to our final frontier: outer space. Within weeks a special committee comprising some of the most influential and respected names in rocketry, spacecraft design and bioastronautics had been formed and tasked with the goal of placing a man into space. Project Mercury had been born.

From the outset, Project Mercury was a fascinating, driven programme, and equally determined men were needed to fly on NASA's spacecraft. They would become known as astronauts – 'star sailors'. However, the business of selecting the first cadre of candidates was a difficult one, not just in determining what relevant qualities were needed in the men but where to find these future space pilots. NASA wanted up to twelve men willing to risk their lives in one of the most hazardous scientific undertakings of all time. The space agency's task was made even more difficult by the fact that no guiding rules or precedents existed for selecting the candidates.

When President Eisenhower gave his approval for the potential space pilots to come from the ranks of serving military test pilots, a mountain of data and medical records was carefully scrutinized. Out of this mammoth undertaking came the names of 110 of the nation's top test pilots. This number then had to be winnowed to just a dozen. The 110 candidates were requested to attend top-secret mass briefings at the Pentagon, where they were given the option of volunteering for a role as a space pilot or simply withdrawing without prejudice. Those who elected to continue were asked to complete initial medical, physical and psychiatric screening. Eventually, 32 names emerged as the best candidates. These elite, highly competitive pilots would now undergo further and far more intense screening to determine which of them would best confront the dangerous unknowns of space and carve for themselves a unique place in history as a Mercury astronaut.

All 32 men had to undergo one of the most rigorous, demeaning and even brutal week-long medical examinations ever undertaken, at the Lovelace clinic in New Mexico. Once completed, another torturous week followed at the Wright Field Aeromedical Laboratory in Ohio, where they were subjected to extreme fitness and physiological testing, the sole purpose of which was to sort out the supermen from the near-supermen. Or, to quote author Tom Wolfe on the subject, they were looking for a group of men with 'The Right Stuff'.

On 9 April 1959, NASA officials introduced seven carefully selected military test pilots as America's first spacemen at a press conference, held in the space agency's headquarters building in Washington, DC. While introducing the men, seated in alphabetical order, an official explained that they would all be given rigorous training, following which one would be selected as the first to ride a rocket into space. At the time, it was hoped that all seven astronauts would have the chance to fly in Project Mercury and later manned space flight programmes.

The seven Mercury astronauts were then named as Lieutenant Malcolm Scott Carpenter, 33 (Navy); Captain Leroy Gordon Cooper Jr, 32 (Air Force); Lt Col. John Glenn Jr, 37 (Marines); Captain Virgil Grissom, 33 (Air Force); Lt Cdr Walter Schirra Jr, 36 (Navy); Lt Cdr Alan B. Shepard Jr, 35 (Navy); and Captain Donald K. Slayton, 35 (Air Force).[1] The selection of the Mercury Seven came at a crucial time in America's history. The fear and uncertainty of the Cold War and the competitive spirit of the evolving Space Race had created international excitement and intrigue in human space flight. All seven men would assist in the engineering development of the Mercury spacecraft, in addition to training for their flights. They were to work for two years from Project Mercury headquarters, then located at Langley Field, Virginia, while the spacecraft they would fly were being refined and tested.

## Primates on the Launch Pads

A 3-kilogram (6½ lb) rhesus monkey named Sam was launched from Wallops Island, Virginia, just before noon on 4 December 1959. Occupying a prototype Mercury capsule, Sam was propelled close to the edge of space to ascertain any possible adverse effects on a human subject. Also positioned inside the capsule were beetle eggs, rat nerve cells, larvae and some bacteria cultures for study under conditions of zero gravity and radiation. The short ballistic flight aboard the Little Joe 2 (LJ-2) carrier rocket was designed to test the spacecraft's escape tower system and how well it functioned under the dynamics of an actual launch. The tripod-shaped tower was essentially an emergency rocket that could be used to separate the spacecraft from the carrier rocket should anything go wrong during the launch phase. The Little Joe rocket was one of a series of boosters designed as a smaller, functional alternative to the Redstone rockets that would one day launch America's first astronauts into space.

After a successful launch, the LJ-2 booster burned out at an altitude of just over 30 km (100,000 ft), at which point the escape tower fired, dragging the capsule away from the spent rocket and propelling it to a peak altitude a little short of 76 km (250,000 ft), during which Sam experienced three minutes of weightlessness. The re-entry was quite mild. Following a ballistic flight lasting 11 minutes and 6 seconds, the capsule splashed down in the Atlantic Ocean under two parachutes and was recovered six hours later by the destroyer USS *Borie*. Once the spacecraft had been hoisted aboard, the hatch was opened and Sam's biopack removed. After a brief examination, Sam was found to be in excellent condition.

Some six weeks later, on 21 January 1960, Sam's mate, Miss Sam, was launched in the fourth test of the Mercury capsule's escape mechanism, aboard another Little Joe rocket (LJ-1B). After splashdown, the Mercury spacecraft was retrieved from the Atlantic Ocean by a Marine helicopter, and three-year-old Miss Sam was found to be unharmed and even frisky following her eight-and-a-half-minute ride into the fringes of space, during which she experienced some 28 seconds of weightlessness.

### Prelude to the First Human Flight

In mid-January 1961, Alan Shepard was selected as the first of the Mercury astronauts to fly into space on a suborbital mission, although his name would not be publicly revealed until the time of his flight. First, however, the Mercury–Redstone rocket combination had to go through a full-on trial, with a chimpanzee substituting for the astronaut in a final check of the Redstone rocket, the Mercury spacecraft, and all of the systems and procedures involved. Everyone at NASA, from the administrator down, had to be fully satisfied that such a trip was survivable.

The chimpanzee chosen for the precursory test flight, which was mostly identical to the one Alan Shepard would fly, was named Ham, one of a group of chimpanzees that had undergone specific training for space flight at Holloman Air Force Base in New Mexico. On 31 January, Ham was wired up, suited up, and strapped into a couch that was inserted into a special capsule. This in turn was gently installed within the Mercury spacecraft, ready for the mission designated MR-2 (Mercury–Redstone 2). His arms were left free, enabling him to push levers when signalled to do so by a sequence of coloured lights, just as he had been trained to do.

After spending several hours waiting on the Cape Canaveral launch pad, Ham was finally launched and on his way. During the ascent he would briefly endure around eighteen times the force of gravity (18 $g$) on his body and travel to a peak altitude of 251 km (156 mi.) at around 9,330 km/h (5,797 mph), a little higher and faster than scheduled. As he soared beyond the atmosphere, Ham experienced a little over six minutes of weightlessness. Then, as the spacecraft slowed on re-entering the atmosphere, he was subjected to a higher than anticipated 14.5 $g$.

At the conclusion of his flight, Ham's spacecraft splashed down 679 km (422 mi.) from Cape Canaveral and began taking on water. It had come down almost 200 km (124 mi.) from the expected landing zone and it took considerable time to determine the spacecraft's where-abouts and dispatch a Marine helicopter. When he reached the splashdown site, the pilot managed to hook onto the spacecraft, now floating on its side and partly submerged. It was then carried to the recovery vessel, USS *Donner*, and gently lowered onto the ship's deck.

Despite his many ordeals, Ham was found to be fine and relatively calm once he had been extracted from his capsule secured within the

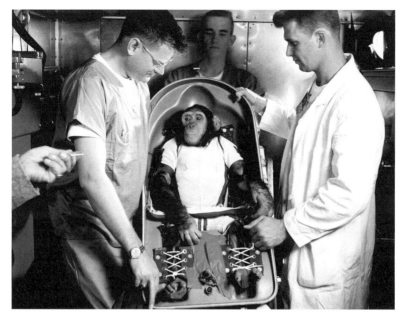

Chimpanzee Ham during training for his MR-2 suborbital flight.

spacecraft. The on-board veterinarian, Richard Benson, determined that Ham's only problems were a little fatigue and dehydration, and a slight bruise to his nose. He devoured an apple handed to him, so his appetite was unaffected. He did, however, grow angry and impatient with all the attention he was receiving, especially the constant snapping and flashing of photographers' cameras; he began baring his teeth and issuing angry screeches.

Altogether, Ham's flight lasted approximately sixteen and a half minutes. Throughout the intense speed, g-forces and weightlessness, he had performed his tasks methodically and well, just as he had been trained to do at Holloman.[2] Ham's days as a space pioneer were now over, and he was transferred to the National Zoo in Washington, DC, in 1963, where he lived for seventeen years, before being transferred to the North Carolina Zoo. He died on 18 January 1983 at an estimated age of 26. His skeleton was preserved and placed on display at the National Museum of Health and Medicine in Silver Spring, Maryland, while the rest of his remains are buried beneath a plaque memorializing him and his space flight at the International Space Hall of Fame, within the New Mexico Museum of Space History in Alamogordo.

Though this was much to the dismay of many, and particularly Alan Shepard, Wernher von Braun decided that problems encountered with a regulator on Ham's Redstone rocket, which had propelled the space-craft higher than expected, forced him to declare that another test flight was necessary. This time, however, there would be no chimpanzee loaded on the flight. No one questioned von Braun's decision, and the perfect launch of the Mercury–Redstone Booster Development (MR-BD) test flight took place on 24 March 1961. By the time a brooding Alan Shepard flew into space six weeks later, a Soviet cosmonaut had already been there and back.

### Meanwhile, in the USSR

In 1958, teams of Soviet officials had begun arriving at all of the major Soviet Air Force stations, asking the commanding officers to name their top pilots, who would later be interviewed for an undisclosed 'special project'. Eventually, 154 potential Air Force candidates were ordered to report to Moscow for top-secret tests and interviews. Following intense screening, that number was finally culled down to twenty. They became the first cadre of Soviet cosmonauts, although their names would be withheld from the public for many years as a closely guarded state secret.

### Hero of the Cosmos

These days, statues of Vladimir Lenin have been toppled and lie strewn across the former Soviet Union. Many others were melted down in factory furnaces. Unlike statues of the once respected Lenin, monuments to one particular hero from the former Soviet Union still stand tall and are lovingly maintained.

Six decades after completing a single orbit of Earth – his only space flight – the name of cosmonaut Yuri Gagarin is still revered across the nation. That largely unheralded feat, a true milestone in history, caused spontaneous and patriotic fervour to erupt like wildfire across the Soviet Union, and his fame transcended the country's difficult conversion from Communism to capitalism.

Gagarin's 108-minute flight aboard a spherical, 2.3-metre-diameter (7.5 ft) spacecraft named Vostok (East) in April 1961 not only scored an impressive first-up victory for the Soviet Union in the superpower space race but secured his place in world history. The son of collective farm workers, and a devoted family man with a renowned sense of humour and a shy, engaging smile, the 27-year-old captured the imagination and

## The First Soviet Cosmonaut Detachment

| Name of Cosmonaut (alphabetical) | Date of Birth | Age at Selection | Date of Official Selection |
|---|---|---|---|
| Anikeyev, Ivan | 12 February 1933 | 27 | 7 March 1960 |
| Belyayev, Pavel | 26 June 1925 | 34 | 28 April 1960 |
| Bondarenko, Valentin | 16 February 1937 | 23 | 28 April 1960 |
| Bykovsky, Valery | 2 August 1934 | 25 | 7 March 1960 |
| Filatyev, Valentin | 21 January 1930 | 30 | 25 March 1960 |
| Gagarin, Yuri | 9 March 1934 | 25 | 7 March 1960 |
| Gorbatko, Viktor | 3 December 1934 | 25 | 7 March 1960 |
| Kartashov, Anatoly | 25 August 1932 | 27 | 28 April 1960 |
| Khrunov, Yevgeny | 10 September 1933 | 26 | 9 March 1960 |
| Komarov, Vladimir | 16 March 1927 | 32 | 7 March 1960 |
| Leonov, Alexei | 30 May 1934 | 25 | 7 March 1960 |
| Nelyubov, Grigori | 31 March 1934* | 25 | 7 March 1960 |
| Nikolayev, Andrian | 5 September 1929 | 30 | 7 March 1960 |
| Popovich, Pavel | 5 October 1930 | 29 | 7 March 1960 |
| Rafikov, Mars | 29 September 1933 | 26 | 28 April 1960 |
| Shonin, Georgi | 3 August 1935 | 24 | 7 March 1960 |
| Titov, Gherman | 11 September 1935 | 24 | 7 March 1960 |
| Varlamov, Valentin | 15 August 1934 | 25 | 28 April 1960 |
| Volynov, Boris | 18 December 1934 | 25 | 7 March 1960 |
| Zaikin, Dmitri | 29 April 1932 | 27 | 25 March 1960 |

*On Nelyubov's grave his date of birth is incorrectly given as 8 April 1934

hearts of a generation both at home and abroad. Meanwhile, his pioneering space flight quickly established that the Soviet Union was well to the fore in space exploration and space politics.

Yuri Alekseyevich Gagarin was born in the village of Klushino, on the eastern edge of the Smolensk region in western Russia, on 9 March

On 12 April 1961 Soviet cosmonaut Yuri Alexeyevich Gagarin became the first human being to be launched into space.

1934. He would be the third of four children born to Alexei Gagarin, a carpenter by trade, and Anna Gagarina. There was no electricity or running water on the local collective farm where the family toiled to make a modest living, but they were a close-knit family. Yuri's mother was a well-educated woman who had grown up in St Petersburg, and she would either read to the children each night or allow them to read for themselves.

Alexei Gagarin had hoped that all three of his sons would become carpenters and one day unite in a family business. The war intervened in those plans, with German troops advancing into the Soviet Union in 1941. The following year, the village of Klushino was overrun by the Nazis, who took savage reprisals against villagers of all ages allegedly involved in any acts of resistance or sabotage. In one such instance, Yuri's younger brother, Boris, was captured and left hanging by the neck from a tree; his mother only just managed to cut him down in time to save his life.

Another incident during the war would have an indelible effect on Yuri Gagarin, when a Soviet aircraft fell from the sky, belching smoke, and crash-landed near Klushino after a one-sided aerial dogfight with a German Messerschmitt. Like other children from the village, Yuri rushed out to the crash site, taking food for the two pilots and helping them to salvage useable parts of the wreckage before they were picked up and rescued by another aircraft, which landed nearby. Yuri, who had clambered all over the downed fighter and talked briefly with the airmen, was in awe of them and their aircraft and began to dream of becoming a combat pilot.

After the war, much to his father's consternation, Yuri decided against carpentry as a career and took on temporary work in a steel foundry in Moscow prior to studying engineering at the Saratov Industrial Technical School. He also joined a local flying club, and once he had completed his first flight in a Yakovlev Yak-18 he knew beyond any doubt what he wanted to do with his life. In 1955, he enrolled as an aviation cadet and quickly became so proficient that his instructor recommended him for the military piloting school at Orenburg. Soon after, at a dance held at the school's air base, he met his future wife, Valentina. They were married in October 1957, and Gagarin later admitted he was too swept up in their wedding preparations to take much notice of dramatic news that month concerning the launch of Sputnik, the world's first Earth-orbiting satellite.

Gagarin was subsequently posted to a Soviet Air Force fighter interceptor squadron located in the bleakness of the Arctic Circle, where, despite the inhospitable surroundings, he worked and flew hard. He and Valentina celebrated the birth of their first daughter, Lena, in April 1959. A few months later, following those top-secret interviews by teams of officials across the Soviet Union, Snr Lieutenant Yuri Gagarin was one of those ordered to report to Moscow as a potential candidate for cosmonaut training. He would be one of the twenty pilots eventually selected as a member of the first Soviet detachment of cosmonauts.[3]

By the early summer of 1960, six of the twenty candidates had been selected as the ones best suited to pilot the first manned Vostok spacecraft (like the first Sputnik, the Soviets would not attach a number to this flight), and it soon became evident that Gagarin was the outstanding candidate. His training intensified even as he and Valentina celebrated the birth of a second daughter, Galya. A week before the flight, set for 12 April, the Soviet State Commission confirmed that Gagarin would be the prime pilot, with Lieutenant Gherman Titov as his backup. In a speech made in the Kazakh town of Akmolinsk on 14 March 1961, Nikita Khrushchev publicly declared that 'the time is not far off' when the Soviet Union would send 'the first spaceship with a man on board . . . into space'.[4] His prediction was realized just four weeks later when Gagarin (twice promoted to major during the flight) became history's first human space traveller.

The launch took place at 9:07 a.m. Moscow time on 12 April, with an elated Gagarin crying out, '*Poyekhali!*' (Here we go!) moments after lift-off. The rocket's second stage shut down on schedule and the Vostok spacecraft slipped into an elliptical orbit with an apogee (high point) of 327 km (203 mi.) and a perigee (low point) of 181 km (112 mi.). No public announcement was made of the event until there was confirmation that the spacecraft had actually achieved orbital status. Fifteen minutes after launch, Gagarin radioed that he was over South America. He drank water and ate a jelly specially prepared by the Soviet Academy of Sciences as the dramatic news of his launch swept across the globe.

By 10:15 a.m., Gagarin was looking down on Africa, stating that 'the flight is normal' and he was comfortable with the state of weightlessness. The landing sequence began soon after. Seventy-nine minutes into the historic flight, the vehicle's retrorockets burned for 40 seconds, slowing it sufficiently to re-enter Earth's atmosphere. At this time, according to plan, the spherical descent module should have separated explosively from the attached equipment module, but a metal cable holding the two components together did not detach. The tethered spacecraft began to spin and tumble erratically, exposing less protected areas of the descent module to the intense heat of re-entry. As the

The Vostok 3K-A rocket on the launch pad ahead of Yuri Gagarin's historic space flight.

temperature inside his spacecraft climbed, Gagarin could only watch helplessly as crimson flames raged around the spacecraft. 'I was in a cloud of fire rushing toward Earth,' he would later recall.[5]

Ten minutes later, the stubborn cable holding the two segments together finally burned through and sheared off with an audible bang. As the descent module continued to fall through the increasingly thicker atmosphere, the wild rotation and swinging gradually dampened, and Gagarin, who had come perilously close to losing consciousness, regained his full senses. Some 7 km (23,000 ft) above the Saratov region of the Soviet Union, the spacecraft's hatch blew off on schedule, and moments later Gagarin was automatically ejected, touching down by parachute near the village of Smelovka. His spacecraft thudded down under its own parachute 3 km (2 mi.) away.[6]

Meanwhile, the Soviet propaganda machine was in full swing, with a dramatic radio announcement bringing news to an elated nation that the cosmonaut had 'safely landed in the prearranged area of the USSR' at 10:55 a.m. after an epic journey lasting 108 minutes, including just over 89 minutes in orbit. Gagarin's Vostok spacecraft had reached a speed exceeding 27,000 km/h (16,777 mph), or about three times faster than any person was known to have flown previously. Following a post-flight debrief session, Gagarin returned to Moscow amid a vast outpouring of jubilation, relief and pride in his accomplishment. He was wildly feted across the Soviet Union and would later embark on an extensive world tour, with crowds eagerly flocking to see and cheer the world's first spaceman. It would be some thirty years before any details of his perilous re-entry were revealed.

### 'Just the First Baby Step'

Inevitably, there are times in a nation's history when its hopes, fears and confidence in its own destiny seem to hinge on the fate of a single person. One of those moments occurred on the spring morning of 5 May 1961, when a 37-year-old U.S. Navy test pilot in Florida squeezed into a cramped Mercury capsule named *Freedom 7*, ready to ride a Redstone rocket into the beckoning skies.

This was a moment that Alan Shepard had relentlessly pursued since his selection as an astronaut two years earlier. Despite this, a hollow feeling dulled his excitement. Whatever accolades he might receive later that day, they would never make up for what he had deemed to be an even greater glory – to be the first person to fly into space. Much to his chagrin, he had fallen just 23 days short of that dream. It was Soviet cosmonaut Yuri Gagarin who had claimed that honour with his single Earth orbit on 12 April.

Alan Bartlett Shepard Jr, who could trace his New England ancestry back through eight generations to the *Mayflower*, was born on 18 November 1923 in Derry, New Hampshire, the son of a banker. He showed an early interest in aeroplanes and took on work at a nearby airfield while still in high school. After graduating from the u.s. Naval Academy in 1944 he saw action in the Pacific aboard the destroyer uss *Cogswell*. Post-war, he gained his aviator's wings and went on to become a skilled, unflappable test pilot before his selection as a NASA astronaut in April 1959.

On the afternoon of 19 January 1961, a day before the inauguration of John F. Kennedy as president, the chief of the Space Task Group, Robert Gilruth, called the seven Mercury astronauts together to confirm who would fly the prized first mission. He wasted little time, announcing that Shepard was the chosen one. There was a stunned silence in the room. 'I did not say anything for about twenty seconds or so,' Shepard later recalled. 'I just looked at the floor. When I looked up, everyone in the room was staring at me. I was excited and happy, of course, but it was not a moment to crow.' The other six men, although deeply disappointed, put smiles on their faces as they congratulated him.[7]

Initially, NASA decided to suppress the name of the first American astronaut so that he could carry out his training without the glare of media attention. Instead, on 22 February 1961, it announced that three candidates had been chosen for final training: John Glenn, Virgil 'Gus' Grissom and Alan Shepard. One was to train as the prime pilot, the other two acting as his backup support. Most people outside NASA felt that the flight would probably fall to the popular and likeable u.s. Marine John Glenn, so when Shepard's name was eventually announced three months later it came as something of a surprise.

Shepard's Mercury–Redstone flight (MR-3) aboard *Freedom 7* was originally scheduled for 24 March, but in late January the Kennedy administration received a damning report on space flight progress from the President's Science Advisory Committee, headed by the president's technical adviser, Jerome B. Wiesner. The committee was recommending a far more circumspect approach in what was seen as unnecessary haste to launch the first astronaut. The Wiesner report cautioned the president that if this haste resulted in the death of an astronaut, it would reflect very poorly on his administration, and called for an immediate delay to the first human space flight, recommending further flights involving primates until it was deemed safe to proceed to a manned shot. Part of the committee's argument was the known unreliability of the Redstone booster. One member of the committee, Dr George Kistiakowsky, even declared that launching Shepard too early would provide the astronaut with 'the most expensive funeral man has ever had'.[8]

The Wiesner report criticized many aspects of NASA's manned space flight programme, placing enormous pressure on Gilruth and the space agency's new administrator, James E. Webb. They discussed the flight at length with key Mercury personnel, later advising Wernher von Braun and his rocket team (with considerable reluctance) that a further unmanned test flight, a so-called 'Mercury–Redstone booster development' launch, would have to be scheduled for the date originally set aside for MR-3. If it proved successful, then Shepard would fly on 25 April. Von Braun, who had already been pressing for another test of the Redstone launch vehicle, was not unhappy with the decision and gave it his approval.

As an impatient Shepard waited, von Braun had his final and trouble-free proving flight. Nineteen days later, a triumphant Soviet Union successfully shot the first man into space. The news both shattered and infuriated Shepard, who had lost the most coveted place in early human space flight history. As the astronauts' nurse, Dee O'Hara, later reflected, 'It was a big blow to everybody and a great disappointment. Gagarin's flight made us all look like fools. Alan was bitterly disappointed, and I could understand that.'[9]

On 2 May, the first attempt to launch *Freedom 7* was postponed 2 hours and 20 minutes before the scheduled lift-off, due to bad weather. A heavy storm had swept through the Cape early that morning, accompanied by lightning and rolling thunder. At 7:25 a.m. Shepard, fully suited up in Hangar S and waiting for news of the delay, received confirmation that the launch was off for that day. The Redstone was already fully fuelled and would have to be drained, and he knew it would be at least 48 hours before they could refuel the rocket and try again. Prior to the postponement, the name of America's first astronaut had been revealed to the public. 'I was relieved when they made the announcement,' he later said. 'It was getting to be a real strain keeping the secret.'[10]

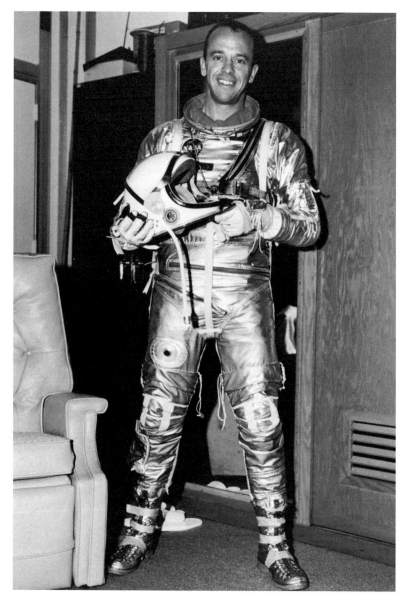

NASA astronaut/U.S. Navy Commander Alan B. Shepard Jr.

Three days later, on 5 May 1961, Shepard was awakened shortly after 1 a.m. Instruments were attached to his body to measure his breathing and heart rate, following which he donned his silvery space suit. As the first faint steaks of dawn began to colour the eastern sky, Shepard stepped down from the transfer van at the launch pad, carrying his portable air-conditioning unit. He took a few steps, then paused for a few seconds to take in the sight of the gleaming rocket that would carry him into space. As he recalled in the Project Mercury astronauts' book, *We Seven*, it was a moment he would never forget:

> I sort of wanted to kick the tires – the way you do with a new car or an airplane. I realized that I would probably never see that missile again. I really enjoy looking at a bird that is getting ready to go. It's a lovely sight. The Redstone with the Mercury capsule and escape tower on top of it is a particularly good-looking combination, long and slender. And this one had a decided air of expectancy about it. It stood there full of lox [liquid oxygen], venting white clouds and rolling frost down the side. In the glow of the searchlights it was really beautiful.[11]

John Glenn and the close-out team were already waiting in the canvas-covered White Room as Shepard approached the open hatch of *Freedom 7*. At 5:20 a.m. he disconnected the hose to his portable air conditioner, slipped off his protective overshoes and made his way feet-first into the spacecraft. Then, suit technician Joe Schmitt strapped Shepard in hard and hooked his suit up to the oxygen system.

At 6:10 a.m., the hatch was closed. 'I was alone,' Shepard later reflected. 'I watched as the latches turned to make sure they were tight.'[12] Seventeen minutes later, the rust-red launch gantry was rolled away from the Redstone, leaving just a yellow cherry-picker with a long boom that would retrieve the astronaut in the case of an emergency. It would remain in position until a minute before the launch.

To everyone's frustration, but mostly Shepard's, there were several delays to the launch. The last came 2 minutes and 40 seconds before launch owing to an abnormally high reading on a fuel pressure regulator. Having endured more than three hours strapped uncomfortably into his spacecraft, which resulted in having to urinate in his special woollen undergarment (rather than exit *Freedom 7* and force another two-day delay), Shepard had finally reached the point of maximum tolerance with all the delays. 'I've been in here more than three hours,' he said, with an edge to his voice. 'I'm a hell of a lot cooler than you guys. Why don't you just fix your little problem and light this candle?'[13]

This latest problem was quickly rectified, and the countdown resumed. With 60 seconds remaining to lift-off, the yellow cherry-picker edged away from the Redstone, leaving the slim rocket poised for lift-off on its pad. The firing command and lift-off came at 9:34 a.m. Moments later, Shepard heard CapCom (capsule communicator) Deke Slayton say, 'You're on your way, José!' – a joking reference to Shepard's favourite comedian, Bill Dana, who had a routine featuring a frightened astronaut named José Jiménez. 'Roger, lift-off,' Shepard calmly responded, 'and the clock has started.'

A minute after leaving the launch pad, the Redstone reached transonic speed. At this stage of the flight the dynamic pressure created as the forces of speed and air density kicked in caused the rocket to begin shaking and vibrating. Shepard was aware that this would happen and simply had to ride it out. Before long, the vibration began to ease, and he was able to report that the ride was a lot smoother.

The Redstone burned out 142 seconds into the flight but continued on its upward trajectory. The escape tower, no longer needed, was jettisoned. Another 38 seconds later, *Freedom 7* was explosively separated from the booster, and it continued its ascent. Shepard continued to monitor his instruments as *Freedom 7* performed a 180-degree turnaround manoeuvre so that its blunt end and Shepard's back would face in the direction of flight, to minimize the re-entry shock. He then initiated a sequence that shut down the capsule's automatic pilot and gave him temporary control over his craft. It was something that Gagarin – always under full automatic control from the ground – had not been able to do. Alan Shepard was no longer a mere observer; as he moved *Freedom 7* through yaw, pitch and roll exercises, he became the first person to actually pilot a spacecraft.

Soon after, *Freedom 7* was back under automatic control once again, allowing Shepard, now weightless, to continue his instrument checks, monitor the spacecraft's fuel and electrical systems and report to Mission Control on his physical condition.

Four and a half minutes into the flight, *Freedom 7* achieved a maximum altitude of 187.5 km (116.5 mi.) as Shepard continued to experience weightlessness, which he later described as 'pleasant and relaxing', despite his heavy work programme on the short mission. Then, all too soon, Deke Slayton was reading out the countdown sequence for retrofire. Just 5 minutes and 14 seconds after he had lifted off on his history-making flight, Shepard fired the retrorockets to slow the spacecraft and positioned *Freedom 7* into the correct attitude for re-entry. The spacecraft arced back into the atmosphere, with Shepard reporting that everything was going smoothly.

The Redstone rocket lifts off the launch pad, 5 May 1961, and Alan Shepard flies into the history books on his MR-3 suborbital mission.

Eight minutes after lifting off, *Freedom 7* was passing through ever denser atmosphere. Within a minute, its speed dropped from 6,803 km/h (4,227 mph) to 549 km/h (341 mph). Meanwhile, Shepard, now flying the spacecraft manually, went from weightlessness to enduring a rapid build-up of g-forces that would reach a maximum of eleven times the normal force of gravity. His centrifuge training had prepared him for this, but he still found great difficulty in trying to communicate with the ground, as his voice was more of a forced grunt.

At 9:44 a.m. a drogue parachute popped out from the top of the plummeting capsule at an altitude of 6.4 km (21,000 ft), which helped stabilize it prior to the main parachute deploying at 10,000 feet. As the spacecraft passed through 7,000 feet, an anxious Shepard felt the main parachute blossom out with a jolt, and he breathed a sigh of relief. A quick glance at his instrument panel also showed that the heat shield's landing bag had dropped down in preparation for softening the splashdown. At 9:45 a.m., now travelling at a sedate 32 km/h (20 mph), the extended landing bag below Shepard hit the water, and the history-making flight of *Freedom 7* was essentially over. The parachute shroud lines were automatically cut loose, and the spacecraft was soon bobbing upright in the Atlantic, ready for the waiting frogmen and Marine helicopters to begin the recovery process.

Once the recovery helicopter had securely hooked onto *Freedom 7*, Shepard removed his helmet, opened the hatch – which he dropped into the ocean according to plan – and exited the spacecraft, sitting on the sill as a 'horse collar' sling was lowered to him, which he slipped over his head before being hoisted into the helicopter. 'Boy, what a ride!' the grinning astronaut remarked to the crew as he was flown to the aircraft carrier USS *Lake Champlain*, 6.5 km (4 mi.) away.[14]

Extensive physical examinations were begun immediately, and Shepard was reported to be in 'excellent' health and suffering no ill effects. His safe recovery was greeted with a mixture of relief and jubilation across the United States. An elated President Kennedy, who had watched the launch live on television in Washington, called Shepard by radio-telephone to offer his congratulations. Other messages poured in, including one to Kennedy the following day from Khrushchev, in which he said that 'this latest outstanding achievement in man's conquest of space opens up unlimited possibilities for the study of nature in the name of progress.'[15] Meanwhile, Shepard's flight was being mockingly derided in the Soviet press, which compared his fifteen-minute ballistic shot to Gagarin's full orbital mission. Nevertheless, Shepard called his flight aboard *Freedom 7* 'just the first baby step aiming for bigger and better things', but it always galled him that an overdose of caution had cost the United States (and him) the chance to be first to enter space.

The flight of Shepard aboard his compact *Freedom 7* spacecraft might seem inconsequential when compared to the complex missions being flown today, but back then it galvanized and united the American people, giving them a renewed sense of pride and achievement. It also set in motion one of the greatest ever scientific undertakings. On 25 May 1961, just twenty days after Shepard's flight, President Kennedy stood before a joint session of Congress and committed the United States to landing a man on the Moon by the end of the decade.

In hindsight, everything about the subsequent race to the Moon between the two competing Cold War superpowers involved a complex history of extremes and superlatives. As Kennedy had so succinctly stated in his historic speech before Congress, 'No single space project in this period will be more impressive to mankind, or more important for the long-range exploration of space; and none will be so difficult or expensive to accomplish.'[16] Perhaps the most notable fact surrounding this is that when the president gave his speech, the nation's manned space flight experience amounted to nothing more than a fifteen-minute suborbital space flight completed just three weeks earlier. Meanwhile, the end of the decade was just eight years away. To many, it seemed an impossible dream.

### Gus Grissom and the Troubled Flight of *Liberty Bell 7*

Two months after Shepard's historic fifteen-minute space flight, NASA was ready to send another of its astronauts into space. For this mission, Air Force captain Virgil 'Gus' Grissom, from the small town of Mitchell, Indiana, would virtually replicate Shepard's mission by flying a similar profile.

On 21 July 1961, Grissom was launched from Cape Canaveral aboard a Mercury spacecraft he had named *Liberty Bell 7*. This followed the example of Shepard, who called his spacecraft *Freedom 7*, not intentionally in recognition of the seven Mercury astronauts but simply because it was numbered as the seventh spacecraft manufactured at the McDonnell Aircraft plant in St Louis, Missouri. However, the other six astronauts liked the fact that, numerically, it also related to them, and the number seven was subsequently included when naming their own Mercury spacecraft.

The 25-metre (83 ft) Redstone rocket and the spacecraft for Grissom's MR-4 mission were altered in minor ways from those used on Shepard's flight. This included the installation of a viewing window which had subsequently been requested by the astronauts. *Liberty Bell 7* would achieve an altitude of just over 190 km (118 mi.) – a little higher than Shepard's *Freedom 7*. Grissom also practised manoeuvring the craft using manual controls, as well as performing other tasks. Following retrofire and re-entry, the drogue parachute was deployed at 30,000 ft, and the main parachute followed, opening and slowing the spacecraft for splashdown at 10,000 ft. *Liberty Bell 7* plunged into the Atlantic at 8:36 a.m., sixteen minutes after Grissom had been lofted into the Florida skies, and fifteen seconds longer than Shepard's earlier mission. Soon after, some unexpected trouble befell Gus Grissom and his spacecraft, and he came very close to dying at the end of his flight.

Once the spacecraft had righted itself in the water, Grissom contacted the pilot of the rescue helicopter, which was already orbiting around *Liberty Bell 7*. He told the pilot, Marine Lieutenant James Lewis, that they could to hook onto the spacecraft in about three minutes. During this time, Grissom went through all his pre-egress tasks, carefully checking the instruments while unhooking himself from the spacecraft. According to procedure, once Lewis had securely latched onto *Liberty Bell 7* the helicopter would raise it a little so that the spacecraft's hatch was well clear of the water. With this accomplished, Grissom would then give the hatch detonation button a solid punch, blowing the hatch outwards and away from the spacecraft. After crawling out, and like Shepard before him, he would then slip into the 'horse collar' rescue

Gus Grissom poses alongside *Liberty Bell 7*.

device lowered by co-pilot John Reinhard. After he had been hoisted up and helped into the helicopter, Lewis would then transport the astronaut and his spacecraft to the nearby waiting aircraft carrier, uss *Randolph*.

While Grissom was engaged in checking the spacecraft's systems, the recovery sequence also dictated that the helicopter crew snip off most of the spacecraft's 4.2-metre (14 ft) whip antenna. It was no longer needed for long-range communications but could interfere with the main rotor when Lewis descended to hook on. This procedure would be carried out by Lieutenant Reinhard, severing the antenna using long-handled metallic shears, which would then allow Lewis to descend and hook onto a sturdy Dacron loop located at the top of the spacecraft. A gentle lift of about half a metre and Grissom could power down *Liberty Bell 7*, disconnect his helmet – thus ending all communications – blow the hatch and slide out onto the sill, where he would slip into the rescue sling and be hauled upwards, together with his spacecraft.

Just as the antenna was being snipped, Grissom heard an unexpected dull thud. Turning, he saw that the hatch had been explosively jettisoned, and seawater was slopping through the open hatchway. Knowing he had very little time to act, he threw off his helmet, grabbed hold of the instrument panel and hoisted himself out of the spacecraft, dropping into the choppy waters of the Atlantic. Swimming away as best he could through the helicopter's rotor wash, he could only watch in disbelief as more and more water gushed through the open hatch and *Liberty Bell 7* began sinking. He noticed that Lewis was having considerable difficulty in hooking onto the Dacron strip and swam over to try and assist. By this time, only the top of the spacecraft was still above water, and the helicopter was having trouble maintaining any height, with its three wheels actually dipping into the sea.

Constantly battered by the rotor wash and whipped-up water, Grissom was more concerned about the recovery of his vehicle than himself. As he flailed about, he realized that he too was sinking, as he had failed to inflate his protective neck dam before evacuating, and water was seeping into his space suit through an open oxygen inlet port. After courageously ensuring the spacecraft was fully hooked up, Grissom gave Reinhard the thumbs up and Lewis poured on as much power as he could muster. While this tremendous battle to lift the near-submerged spacecraft and the water it now contained was going on, the helicopter moved away from Grissom, along with the dangling rescue sling.

Fortunately, Lewis had thought to contact the backup helicopter and asked Captain Phillip Upshulte to retrieve Grissom from the water, as he was fully occupied in trying to haul the massively overweight spacecraft out of the water. Then, a chip detector warning light lit up on Lewis's instrument panel – an ominous indication that there were metal filings in the oil system. He knew that once the filings spread throughout the engine it could fail within minutes, and the stricken helicopter would quickly follow its leaden cargo into the sea.

Grissom was now finding it difficult to keep his head above the turbulent surface, swallowing seawater and battling physical exhaustion as his waterlogged suit kept dragging him down. He just managed to grab hold of the rescue sling from the second helicopter as it reached him and struggled into the circular collar, facing backwards. With water streaming from his space suit, he was hauled up into the open door of Upshulte's helicopter. Once he was safely on board he shed the sling and shook rescuer Lieutenant George Cox's hand in gratitude, saying a heartfelt, 'Boy, am I glad to see you!'

Meanwhile, the battle to retrieve *Liberty Bell 7* had been lost. Not wishing to ditch in the icy Atlantic, Lewis had reluctantly cut the spacecraft loose. It quickly sank beneath the waves. Once he was on board the *Randolph*, Grissom began his post-flight medical check and debriefing. Following an initial examination by physicians, he was proclaimed to be in good health, despite having come close to drowning. President Kennedy had watched the entire flight by television in the White House, and he contacted the astronaut by radio-telephone to congratulate him and express satisfaction at his successful flight. After an overnight stay and further medical and psychological checks at Grand Bahama Island, Grissom was flown back to Cape Canaveral, where he was presented with the NASA Distinguished Service Medal at a press conference.

In spite of Grissom's valiant efforts to save the sinking spacecraft, which nearly cost him his life, there were those who questioned whether he had accidentally (or even purposely in panic) blown the hatch. NASA would have none of it, and vindication came when he was later selected to command the first Gemini two-man space mission, and the first Apollo mission. Tragically, Grissom and fellow crew members Ed White and Roger Chaffee would die in January 1967 during a pre-flight test, when an oxygen-fuelled fire gutted the interior of their Apollo spacecraft as it sat atop an unfuelled Saturn I launch vehicle.

## Manned Suborbital Mercury–Redstone Flights

| Mission | Astronaut | Launch Date | Maximum Altitude | Duration |
|---------|-----------|-------------|------------------|----------|
| MR-4 | Alan B. Shepard Jr | 5 May 1961 | 187.50 km (116.51 mi.) | 15 mins 22 secs |
| MR-3 | Virgil I. 'Gus' Grissom | 21 July 1961 | 190.32 km (118.26 mi.) | 15 mins 37 secs |

Had he lived, Gus Grissom would have been delighted to know that *Liberty Bell 7* was finally located on 1 May 1999, at a depth of nearly 4.8 km (3 mi.) in the Atlantic Ocean, northwest of Grand Turk Island. The Discovery Channel had financed the search expedition, led by undersea salvager Curt Newport, who had previously worked on the recovery of space shuttle *Challenger* and twa Flight 800, lost in the Atlantic off New York in 1996. At 2:15 a.m on 20 July 1999, 38 years after it sank and 30 years to the day after the first two men walked on the Moon, *Liberty Bell 7* was hoisted from the floor of the Atlantic Ocean to the surface.[17] Following the spacecraft's recovery, it was inspected, completely dismantled and thoroughly cleaned, and some parts were replaced before it was reassembled, ready to be displayed in its new home at the Cosmosphere in Hutchinson, Kansas.

Unfortunately the spacecraft hatch was never recovered. Tests were carried out, and it was subsequently reported that one likely cause of the hatch being blown was a static discharge occurring at the same moment the metal antenna atop *Liberty Bell 7* was being cut.

## A Winged Spacecraft

Even as nasa was ramping up preparations to launch the first astronauts into space, another programme was under way that would involve a select group of aerospace pilots reaching ever higher into the atmosphere, eventually flying beyond the boundaries of space. The vehicle that would take them there, meanwhile setting new aviation speed records, was the mighty winged spaceplane, the North American x-15.

Aerospace historian Michelle Evans, author of *The x-15 Rocket Plane: Flying the First Wings into Space*, has described the hypersonic rocket-powered aeroplane as 'the most successful research aircraft ever flown'. As Evans states, 'Twelve pilots flew 199 missions from June 1959 until October 1968. They reached Mach 6.7 (4,520 mph) and an altitude of 354,200 feet (67.1 mi.) during the course of the programme. Eight of the twelve pilots flew high enough to earn astronaut wings (although only five of them actually were awarded during the programme). One pilot lost his life.'¹⁸ Altogether, the x-15 programme used three of its rocket planes to fly to the then-record height quoted by Evans and as fast as nearly seven times the speed of sound. Volumes of test data gleaned from the 199 missions of the x-15 greatly assisted in shaping NASA's successful Mercury, Gemini, Apollo and space shuttle human space flight programmes.

On 8 June 1959, aeronautical engineers began a series of free-flight tests with the first of the three experimental x-15 rocket planes, a research aircraft designed to help provide data on high-altitude flight at the fringes of space – amid hopes of one day exceeding that elusive boundary in the sky. The sleek black aircraft had been built by North American Aviation, Inc. (NAA) under contract to the U.S. Air Force. That day, the $5-million aircraft was dropped from beneath the wing of a specially modified B-52 bomber, 11.5 km (38,000 ft) above the Mojave Desert. This was the first unpowered glide test, flown by 37-year-old NAA civilian test pilot Scott Crossfield, who successfully guided the x-15 to a landing four minutes later on the Rogers Dry Lake runway, attached to Edwards Air Force Base in California. The x-15 had previously been carried aloft four times on 'captive' flight tests under the B-52's right wing.

The second of three eventual x-15 aircraft successfully completed its first two rocket-powered flights on 17 September and 17 October 1959, again with Crossfield at the controls. On both flights it exceeded 2,250 km/h (1,400 mph), and on the second powered flight Crossfield

The North American x-15 rocket plane in flight.

had taken the x-15 to 18.3 km (60,000 ft) in altitude. On the third powered flight, on 5 November, there had been a fuel explosion in the lower of the two engines that damaged the aircraft, but Crossfield managed to nurse it back to Edwards and completed a safe emergency landing. The x-15 programme continued, along the way breaking many existing speed and altitude records.

On 27 June 1962, NASA's chief research pilot, Joseph A. 'Joe' Walker, set a world speed record for a piloted winged aircraft of 6,605 km/h (4,104 mph) in x-15 No. 1. Meanwhile, one of the goals of the programme was to fly beyond what was officially recognized by the U.S. Air Force as the boundary of space. This boundary has been the subject of much conjecture and disagreement for many decades. Last century, the Hungarian-born physicist Theodore von Kármán declared it to be 50 mi. (close to 80 km) above sea level. However, the governing body for registering aeronautic and astronautic records, the Fédération Aéronautique Internationale (FAI), and the National Oceanic and Atmospheric Administration (NOAA) define the boundary of space as 100 km or 62 mi. altitude. Yet NASA, the Federal Aviation Administration (FAA) and the United States Air Force (USAF) still regard the Kármán line as representing where space truly begins. As a consequence, the Air Force will award astronaut wings to any pilot who flies higher than 50 miles.

The following month, on 17 July 1962, Air Force Major Robert White was launched from beneath the wing of the B-52 some 13.7 km (45,000 ft) over Delamar Lake, Nevada, and flew x-15 No. 3 to an altitude of 95.92 km (59.6 mi.), marking the first time a pilot had flown into the new realm of space. As a result, White became the first x-15 pilot and the fifth American to qualify as an astronaut (a goal already achieved by NASA astronauts Alan Shepard, Gus Grissom, John Glenn and Scott Carpenter). The next day at the White House, President Kennedy presented Walker, White and two former x-15 pilots, Scott Crossfield and Forrest Petersen, with the 1961 Robert J. Collier Trophy for outstanding contributions to aeronautics and astronautics.

On 17 January 1963, marking the 77th flight of the x-15, Joe Walker was able to push the mighty research aircraft to an altitude of 82.7 km (51.4 mi.), just over the Air Force's declared boundary of space. At the time, Walker was told he did not qualify for astronaut wings, as only the military had astronaut wings to confer on their pilots, and he was regarded as a civilian pilot. This anomaly existed until 23 August 2005, when three civilians from the total of eight x-15 pilots who had flown

## The X-15 Pilots and Their Records
### (USAF-recognized boundary of space is 50 mi./80 km above sea level)

| Pilot | Number of Flights | USAF Space Flights | Maximum Speed | Maximum Altitude |
|---|---|---|---|---|
| Michael Adams, USAF | 7 | 1 | 6,150.9 km/h (3,822 mph) | 80.95 km (50.3 mi.) |
| Neil Armstrong, NASA | 7 | 0 | 6,419.7 km/h (3,989 mph) | 63.09 km (39.2 mi.) |
| Scott Crossfield, NAA | 14 | 0 | 3,152.7 km/h (1,959 mph) | 24.62 km (15.3 mi.) |
| William Dana, NASA | 16 | 2 | 6,271.6 km/h (3,897 mph) | 93.50 km (58.1 mi.) |
| Joe Engle, USAF | 16 | 3 | 6,255.5 km/h (3,887 mph) | 85.46 km (53.1 mi.) |
| William Knight, USAF | 16 | 1 | 7,272.6 km/h (4,519 mph) | 85.46 km (53.1 mi.) |
| John McKay, NASA | 29 | 1 | 6,216.9 km/h 3,863 mph) | 89.96 km (55.9 mi.) |
| Forrest Petersen, USN | 5 | 0 | 5,793.6 km/h (3,600 mph) | 30.90 km (19.2 mi.) |
| Robert Rushworth, USAF | 24 | 1 | 6,464.7 km/h 4,017 mph) | 86.74 km (53.9 mi.) |
| Milton Thompson, NASA | 14 | 0 | 5,991.6 km/h (3,723 mph) | 65.18 km (40.5 mi.) |
| Joseph Walker, NASA | 25 | 3 | 6,604.7 km/h (4,104 mph) | 107.83 km (67.0 mi.) |
| Robert White, USAF | 16 | 1 | 6,585.4 km/h (4,092 mph) | 95.92 km (59.6 mi.) |

NAA = North American Aviation; NASA = National Aeronautics and Space Administration; USAF = United States Air Force; USN = United States Navy

beyond the 50-mile mark, namely Joe Walker, William H. 'Bill' Dana and John B. McKay, were officially recognized for their achievements and awarded astronaut wings, although by this time both Walker and McKay were deceased.[19]

Air Force Major Michael J. Adams had joined the X-15 programme on 20 July 1966 and would make a total of seven flights in the No. 1 and No. 3 aircraft. On 15 November 1967, while on flight 191 of the programme in X-15 No. 3, he had breached the required altitude for space flight. When he began his re-entry into the atmosphere, he got into control difficulties and the X-15 entered into a violent high-speed spin and began to break up under excessive aerodynamic stresses. The

## X-15 Flights Exceeding 50 mi./80 km Altitude

| Flight No. | Pilot | Date | Altitude |
|:---:|:---:|:---:|:---:|
| 62 | Robert White | 17 July 1962 | 95.92 km (59.6 mi.) |
| 77 | Joseph Walker | 17 January 1963 | 82.72 km (51.4 mi.) |
| 87 | Robert Rushworth | 26 June 1963 | 86.74 km (53.9 mi.) |
| 90 | Joseph Walker | 19 July 1963 | 105.89 km (65.8 mi.) |
| 91 | Joseph Walker | 22 August 1963 | 107.83 km (67.0 mi.) |
| 138 | Joe Engle | 29 June 1965 | 85.46 km (53.1 mi.) |
| 143 | Joe Engle | 10 August 1965 | 82.56 km (51.3 mi.) |
| 150 | John McKay | 28 September 1965 | 89.96 km (55.9 mi.) |
| 153 | Joe Engle | 14 October 1965 | 81.11 km (50.4 mi.) |
| 174 | William Dana | 1 November 1966 | 93.50 km (58.1 mi.) |
| 190 | William Knight | 17 October 1967 | 85.46 km (53.1 mi.) |
| 191 | Michael Adams* | 15 November 1967 | 80.95 km (50.3 mi.) |
| 197 | William Dana | 21 August 1968 | 81.43 km (50.6 mi.) |

*Killed in crash of x-15-3

disintegrating aircraft slammed into the ground northwest of Randsburg, California, instantly killing Adams.

Towards the end of 1968, the record-breaking flights of the x-15 were all but forgotten as the attention of the world turned instead to the final months of President Kennedy's pledge to place men on the Moon by the end of the decade, and the forthcoming flight of Apollo 8, which would successfully orbit the Moon on the first crewed lunar mission.

On 24 October that year, Bill Dana made a perfect landing on Rogers Dry Lake at Edwards, the 199th flight involving the three x-15 research aircraft. There were plans to round out the programme with a 200th flight, but following a number of cancellations and aborts it was decided the programme's time had come. Interestingly enough, two of the x-15 pilots, Neil Armstrong and Joe Engle, would go on to be selected as NASA astronauts. Armstrong, who made seven flights in the x-15, later become the first man to set foot on the Moon, in Apollo 11. Meanwhile Joe Engle – sadly bumped from the Apollo 17 mission to make way for trained geologist Harrison Schmitt – would go on to fly many of the Approach and Landing Tests (ALT) in the space shuttle programme. He would later be selected to command the second orbital test mission, STS-2, thus becoming the only person to fly both the X-15 and the space shuttle.

# 3

# INTO ORBIT

In early planning for the first Mercury–Redstone flights, NASA had considered launching each of its seven astronauts on several suborbital test flights before venturing into Earth-orbital missions using the far more powerful Atlas rocket. Eight Redstone rockets had been purchased, so this was achievable, but subsequent events caused those plans to be abandoned.

On 21 November 1960 the Redstone used on the first Mercury–Redstone flight (MR-1) was badly damaged in a launch anomaly that came to be known as the 'four-inch flight'. Lift-off came right on schedule, but just as the Redstone was about to leave the launch pad, its engine suddenly shut down. The Redstone wobbled, then settled back on its fins. The rocket-propelled escape tower attached to the top of the Mercury capsule self-activated and soared high into the sky, following which the drogue parachute popped out of the top of the Mercury spacecraft, still firmly attached to the top of the Redstone. This was followed moments later by the main and reserve parachutes, which fluttered down to the ground beside the reluctant rocket, still teetering unsteadily on its launch pedestal. It was estimated that the Redstone had achieved only a 4-inch lift-off, later traced to an electrical fault, but it was enough to render the rocket unusable.[1]

The following month, another Redstone completed the earlier failed flight on a mission designated MR-1A. On MR-2, the chimpanzee Ham completed his successful suborbital flight in a Mercury capsule on 31 January 1961. With concerns raised about the performance of the Redstone booster on that flight, another test launch, designated the Mercury–Redstone Booster Development flight (MRBD), was inserted into the Mercury–Redstone programme, and then two more Redstones were expended on the suborbital flights of Shepard and Grissom.

As both crewed suborbital missions were successfully completed – apart from the loss of the MR-4 spacecraft *Liberty Bell 7* – and with pressure now being applied to NASA following the orbital flight of Yuri Gagarin, it was decided to terminate the Mercury–Redstone programme early and move on to orbital missions with the Atlas rocket.

Knowing of NASA's plans to commence manned orbital flights well in advance, Nikita Khrushchev directed the Chief Designer (Sergei Korolev) to eclipse whatever the Americans proposed. At first, the follow-up flight of Vostok-2 was planned as a three-orbit mission, but Khrushchev did not merely want to emulate NASA's forthcoming mission. To increase the flight's propaganda value and continue demonstrating superior space flight technology, he demanded that it last an entire day.

At 9:00 a.m. Moscow time on 6 August 1961, Vostok-2 was launched in the nose of a multi-stage Vostok-K rocket from the Baikonur cosmodrome, with 26-year-old Soviet Air Force Major Gherman Titov on board. He had already completed his first orbit of Earth before news of this latest manned flight was broadcast over Radio Moscow. Titov carried out routine observations, manoeuvred his Vostok craft twice

Cosmonaut Gherman Titov: the first person to spend a day in space.

using manual controls, ate three meals in paste form from tubes, and managed to sleep for just over eight hours as his flight continued. Unfortunately, he was also afflicted with a mysterious nauseating malady that no one could explain. It would be some years before this illness was identified as space adaptation syndrome (sas), which would strike many spacefarers despite all precautions being taken. Even today, it is difficult to predict who will fall victim to sas. Most space travellers are apparently unaffected, but for others the symptoms can range from mild headaches through to extreme nausea and prolonged vomiting, leaving them malnourished and dehydrated. While the illness is similar to terrestrial motion sickness, scientists believe that sensors in the inner ear trigger a signal to the brain once weightlessness is achieved, so astronauts are warned against making any sudden head movements or playful body motions. Generally, the symptoms will ease after about three days in orbit.

Otherwise, Titov's flight was progressing well. Vostok-2's orbit was very close to that which had been planned, and the spacecraft was circling the Earth once every 88.6 minutes at an average speed of 28,000 km/h (17,300 mph). After completing seventeen orbits, the braking rockets were fired, and Vostok-2 re-entered the atmosphere, landing at Krasny Kut, 64 km (40 mi.) east of Saratov on the Volga River. Three days later, Titov would reveal at a press conference that he had ejected himself from Vostok-2 at 6,500 m (21,325 ft) after his 25-hour flight and had landed by parachute.[2]

At the same press conference, Vladimir Yazdovsky, an expert in space medicine, disclosed that Titov had suffered 'unpleasant sensations' while in weightlessness. Two months later, Yazdovsky and Oleg Makarov (a later cosmonaut), both members of the Soviet Academy of Sciences, disclosed in a scientific paper delivered before the 11th annual congress of the International Astronautical Federation in Washington, DC, that the cosmonaut had been ill for 'the considerable portion of the flight'. This malady had caused 'a disturbance of spatial analysis' (a feeling of disorientation) and loss of balance, but was not enough to prevent Titov from 'a sufficient level of working capacities'.[3]

Though he was lauded across the Soviet Union for this latest space spectacular, Titov's bout of sas brought about fears of some unknown fault in his physiology that was affected by prolonged weightlessness, and he was quietly banned from making any further flights into space. One record still maintained to this day by Gherman Titov is that, at

26 years of age when he flew the Vostok-2 mission, he remains the youngest person ever to fly into space.

## NASA Responds

Born in Cambridge, Ohio, on 18 July 1921, John Herschel Glenn Jr was overwhelmingly regarded as the most popular and personable of the seven Mercury astronauts. When they were first introduced to the American public at their 9 April 1959 press conference, six of the men looked distinctly ill at ease, and their monosyllabic responses to questions posed from the floor of the auditorium were cautious, hesitant and hardly the stuff of headlines. It got even worse when one of the questions regarded how their wives and children had reacted to their selection as astronauts. Some mumbled that their families were pleased and happy. When the relaxed, smiling, freckle-faced Marine test pilot spoke up, he instantly became a PR dream for NASA. He had already achieved a degree of fame in 1957, creating a new trans-American speed record in a supersonic Vought F8U Crusader jet during Project Bullet, a test of the sustained capability of the F8U at near maximum power on a non-stop flight from the west coast to the east coast, which entailed three aerial refuellings. The project name came from the fact that the Crusader could fly faster than a .45 calibre bullet. He had even appeared on the TV show *Name That Tune* alongside child star Eddie Hodges (who later recorded the hit song 'I'm Gonna Knock on Your Door'), on the same day the Soviet Union launched the first Sputnik into space.[4] It is worth recording Glenn's spontaneous response to reporters at the press conference:

> I got on this project because it probably would be the nearest to heaven I will ever get and I wanted to make the most of it [laughter from the audience]. But my feelings are that this whole project with regard to space sort of stands with us now as, if you want to look at it one way, like the Wright Brothers stood at Kitty Hawk about fifty years ago, with Orville and Wilbur pitching a coin to see who was going to shove the other one off of the hill down there. I think we stand on the verge of something as big and as expansive as that was, fifty years ago . . . I think we are very fortunate that we have, should we say, been blessed with the talents that have been picked for

Mercury astronaut/U.S. Marine Corps Lt Col. John H. Glenn Jr.

something like this. I think we would be almost remiss in our duty . . . if we didn't make the fullest use of our talents in volunteering for something that is as important as this is to our country and the world in general right now.[5]

From that time on, the American public fell in love with the handsome, patriotic Marine, and there was universal surprise when he was not selected over Alan Shepard to make the first American space flight. Nevertheless, he would achieve everlasting fame for making the first orbital flight by a NASA astronaut, on a flight filled from the outset with the greatest of expectations and, later in the mission, with high human drama.

An unmanned Mercury spacecraft was launched into a single Earth orbit from Cape Canaveral on 13 September 1961, carrying what was referred to as an 'artificial astronaut'. The flight was designed to demonstrate the ability of the Atlas carrier rocket to propel the Mercury capsule into a satisfactory orbit, as well as to assess the capability of the capsule and its systems to operate completely autonomously. It would also provide a crucial test of the Mercury Tracking Network, while the payload comprised a pilot simulator (a test of the environmental controls), a life support system, and instrumentation that would record and monitor the levels of noise, vibration and radiation. The mission, designated Mercury–Atlas 4 (MA-4), completed a single Earth orbit before re-entering the atmosphere. Having achieved a highly successful mission, the spacecraft was plucked from the ocean by the destroyer USS *Decatur*, 260 km (162 mi.) east of Bermuda.

As Shepard and Grissom had only been weightless for five minutes on their respective suborbital flights, they had been given very little chance to explore this phenomenon. With the reassurance of the MA-4 mission behind them, NASA now wanted to ensure that an astronaut would be able to perform some simple tasks while weightless over an extended period. Scientists were still expressing legitimate concerns that an astronaut, on seeing the planet passing rapidly beneath them, might lose their normal concept of up and down, speed and direction, and become critically disoriented. NASA decided the question would be resolved by conducting a precursory orbital flight using a suitable chimpanzee from those undergoing training with the 6571st Aeromedical Research Laboratory at Holloman Air Force Base. Out of the available candidates, a chimp named Enos was considered to be the most intelligent of all the trained primates and was selected for the mission.

Meanwhile, on 4 October 1961 – coincidentally the fourth anniversary of the launch of the first Sputnik satellite – John Glenn was appointed to fly the first three-orbit mission, with Scott Carpenter to function as his backup pilot. As Glenn would later write of his upcoming mission:

> From a technical standpoint, the orbital mission would be quite different from the ballistic flights in several respects. For one thing, we would be using the Atlas missile as a launch vehicle in order to get up the required thrust and velocity to get into orbit. The Atlas' engines have a total thrust of 360,000 pounds [163,300 kg] compared to 76,000 [34,500 kg] for the Redstone, and the Atlas would get the capsule up to a top speed of nearly 18,000 miles per hour, which is more than three times the speed of the Redstone IRBM, which had served us well on the ballistic flights. Once in orbit, the flight would also last longer – about four and a half hours if we made all three orbits. This would be a new magnitude of space flight for the U.S., and if it were successful it would pave the way for longer voyages, eventually to the Moon and beyond, just as Al and Gus had paved the way for this one. It would be a prelude to our plans for the future.[6]

The precursory flight that would carry Enos into orbit, designated MA-5, was scheduled to complete three orbits of Earth, the same as would be attempted on Glenn's subsequent flight. On 29 November, five hours before launch, the specially constructed primate couch in which Enos had been secured was inserted into the Mercury spacecraft. Following a series of delays, the Atlas-D rocket launched at 10:08 a.m. EDT (Eastern Daylight Time), delivering nearly five times the lift-off power Shepard and Grissom had experienced.

Enos did not seem to be overly concerned by the g forces he encountered during launch, or the subsequent weightlessness. During his second orbit, the lever on the motor skills test malfunctioned and, much to his annoyance, Enos began receiving small electrical shocks through his feet, which in training was meant to indicate he had made an error, rather than being rewarded with a banana pellet for each correct response. He was well trained, however, and despite this ongoing discomfort he continued pulling the levers in sequence. Then his suit began overheating and the automatic attitude controls malfunctioned, causing the spacecraft to repeatedly roll some 45 degrees before thrusters fired to correct the vehicle's orientation. Given these problems, Mission Control decided to end the flight early. Altogether, Enos would experience 181 minutes of weightlessness, without any evident side effects. Three hours and 21 minutes after lift-off, the Mercury spacecraft re-entered Earth's atmosphere and splashed down in the Atlantic, south of Bermuda. Only 75 minutes later, the spacecraft had been located and was hauled aboard the recovery ship, uss *Stormes*.[7]

The spacecraft had functioned well, as had Enos, although he managed to rip loose many of his irritating monitoring sensors, as well as his urinary catheter, while waiting to be recovered from the ocean. His appetite was obviously unaffected by his experience of space flight, as he eagerly consumed an apple and half an orange after he had been removed from the spacecraft and released from all his restraints. Once back on dry land, the unfazed chimpanzee was hailed as the newest hero of the Space Age by media representatives attending a NASA press conference.

Thanks to the successful MA-5 mission and the stoic performance of Enos, NASA concluded that an astronaut could safely withstand orbital space travel and comfortably work under conditions of weightlessness. John Glenn's MA-6 orbital mission could proceed.

### Three Orbits and a Safe Splashdown

The patience of NASA, John Glenn and the American public would be severely tested as 1961 drew to a close. Originally scheduled to be launched on 20 December, Glenn's three-orbit Mercury flight would experience numerous frustrating delays, caused by bad weather and technical issues with the Atlas launch vehicle (designated 109-D) and systems within the spacecraft, which Glenn had earlier named *Friendship 7*. On 27 January 1962, poor weather conditions forced yet another postponement after Glenn had spent 5 hours and 13 minutes inside his spacecraft. Then, on 15 February, NASA called off its ninth scheduled launch attempt as storms continued to lash the Atlantic landing site, following which any further attempts were placed on a day-to-day basis. Three days later, the launch was delayed for a tenth

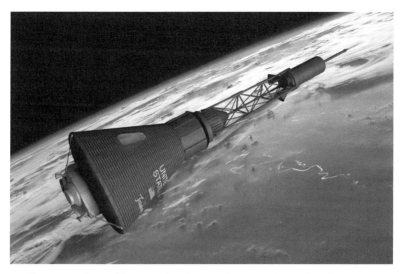

An illustration of *Friendship 7* in orbit, showing the retrorocket pack and restraining straps. Although it would have been explosively jettisoned during the launch phase the red escape tower is shown still in place to demonstrate how it was attached to the spacecraft.

time, and strong rumours began to spread that the flight might be called off for another month. By this time, however, the bad weather had begun to break, so another attempt was scheduled for 20 February. Two months after the original launch date, early on the morning of 20 February, John Glenn was once again strapped into *Friendship 7*, and his oft-delayed journey into space would finally begin.

At 9:47 a.m., Atlas 109-D began belching smoke and fire on Launch Pad 14 at Cape Canaveral. As the engines built to their full thrust, the booster remained fast on the pad until large hold-down clamps dropped away, unleashing the Atlas. Inside *Friendship 7*, Glenn experienced an exhilarating surge as the booster climbed into the Florida skies, before the automatic guidance system built into the Atlas initiated a roll onto a scheduled northeast heading. Approximately 45 seconds after launch, Glenn entered into the so-called 'High-Q' zone – the area of the flight where the booster and spacecraft would encounter the highest aerodynamic forces acting against them. Over the next 30 seconds he was subjected to some severe buffeting, but this soon eased as the air outside thinned.

Two minutes and 14 seconds after lift-off, the two big outboard booster rockets shut down and were discarded, leaving just the central (or sustainer) engine to propel *Friendship 7* into orbit. Twenty seconds later, right on schedule, the red escape tower above the spacecraft was also jettisoned, belching out a cloud of flame and smoke. Next, again perfectly on schedule, the sustainer engine cut off, following which explosive bolts fired, separating the booster and capsule. Posigrade rockets then fired to push the spacecraft away from the spent booster. Glenn's periscope extended and *Friendship 7* began to pitch over to the position it would maintain throughout the three orbits, blunt end first.[8] As this happened, Glenn, now weightless, was able to gain his first view of Earth's horizon. He could see for hundreds of kilometres in every direction and reported in excitement, 'Zero G and I feel fine; capsule is turning around. Oh, that view is tremendous!'[9]

During the flight, the automatic control system didn't respond as programmed, causing the spacecraft to swing over to one side along the yaw axis and then correct itself with a large expenditure of hydrogen peroxide propellant, normally released in small pulses as required. This meant Glenn had to take over manual control. The same problem would occur again, with *Friendship 7* this time yawing in the opposite direction. For most of the rest of his flight, Glenn controlled the craft himself.

Shortly after completing his first orbit, Glenn reported seeing thousands of luminous particles surrounding *Friendship 7*, which he compared to a swarm of fireflies. He could create even more by banging on the side of his spacecraft. They were later identified on Scott Carpenter's subsequent three-orbit mission as simply small flakes of frost from the side of the craft lit by the intense sunlight.

There were some anxious moments on the third and final orbit when a faulty indicator gave an erroneous signal that Glenn's heat shield had become dislodged, a potentially life-threatening situation that could have incinerated him and his spacecraft upon re-entry. Ground controllers decided not to jettison the retrorocket package as planned, hoping the straps that held it in place would also stabilize the possibly errant heat shield. Walter Schirra, at the tracking station in California, passed this information to Glenn. As this was not according to routine, Glenn asked for the reason, but Schirra said that the next tracking station, in Texas, would explain.

Glenn could not spend much time pondering the situation, as he was preparing *Friendship 7* for re-entry, placing it in the correct attitude for retrofiring. It was only then that he was told that Mercury Control had received an earlier indication by telemetry that the heat shield might have come loose. The shield, made up of a thick coating of resinous material on the blunt face of the spacecraft, was designed to dissipate the heat and energy encountered during re-entry by melting and boiling away very slowly. Glenn was all too aware that only the heat shield stood between him and disaster as he encountered the intense heat of atmospheric re-entry. If the shield was indeed loose, then the retro-package and its retaining straps would have to hold it in place – a forlorn expectation.

The re-entry was anything but routine. The outside heat quickly built up in intensity and Glenn saw one of the three metal retro-straps flapping around in front of his window before it burned off. Flaming chunks of the retrorocket package then flashed by. 'That was a real fireball!' Glenn exclaimed. Fortunately, the heat shield held, and Glenn splashed down uneventfully in the Atlantic about 65 km (40 mi.) short of the predicted area.

The crew of the destroyer uss *Noa* winched the astronaut aboard some 37 minutes later, still secured inside his capsule. He then warned the ship's crew to stand clear, and hit the hatch detonator plunger with the back of his hand. The plunger recoiled, cutting his knuckles slightly through his gloves, thus inflicting the only injury he received during the mission.

Lieutenant Commander Robert Mulin and Army physician Gene McIver conducted a preliminary examination aboard the *Noa*, later describing Glenn as being hot, sweating profusely, fatigued and a little dehydrated. Following a glass of water and a shower, the world's newest spaceman quickly recovered, despite having lost 2.41 kg (5 lb 5 oz) from his pre-flight weight. At a general physical examination conducted later at Grand Turk Island, physicians found very few measurable differences between the pre-flight and post-flight medical findings, leading physicians to conclude that space did not seem to be quite as inhospitable an environment as first feared.[10] A post-flight investigation revealed that the heat shield had, in fact, been quite secure; the problem was traced to a faulty sensor.

Glenn was not to know it at the time, but in the space of the five hours of his three-orbit flight he had become an instant national hero, lauded in the media as the embodiment of America. His grinning, freckled face appeared on countless front pages of newspapers and magazines. He was decorated at the White House by President Kennedy, addressed a specially convened session of Congress, and was honoured with massive, exultant parades throughout the United States in scenes of adulation not seen since the triumphant celebrations of Charles Lindbergh's historic transatlantic flight 34 years earlier.

## The 'Lost' Astronaut

Measured in time, it was only 35 minutes, but across the United States, sombre news bulletins were preparing people for the worst possible news – that the nation's newest astronaut, Scott Carpenter, had perished at the end of his three-orbit space flight. Then came a very welcome announcement from NASA's public affairs officer, John 'Shorty' Powers: 'A Navy P2V Neptune has reported sighting the spacecraft floating in the landing area. Alongside it was a life raft, and sitting in it was a gentleman named Carpenter.' A massive sense of relief swept the nation.

It was 24 May 1962, and that day's adventure had begun somewhat explosively for Lieutenant Commander Scott Carpenter when the Atlas 107D rocket carrying his *Aurora 7* spacecraft had lifted off from Cape Canaveral at 8:45 a.m. EDT, tearing a path into the Florida skies to begin a three-orbit flight that would mostly emulate that of John Glenn three months earlier. During his first orbit, Carpenter reported an overheating problem with his space suit, and that his attitude indicator instruments were at odds with his visual estimate. Additionally, and quite concerning to those in Mercury Control, he reported that the level of hydrogen peroxide propellant – used to control the spacecraft's attitude – had diminished to 69 per cent of capacity. He was therefore directed to manually control his craft's attitude in order to conserve fuel.

On his second orbit, Carpenter released a multicoloured balloon, which trailed behind *Aurora 7* as an experiment in determining which colour was best for visual observation. Although the balloon did not fully inflate, Carpenter did say that orange stood out most, with silver also easy to distinguish. He then reported that his space suit was heating up again, but he was able to bring the temperature down to a comfortable level. Mission Control soon transmitted the news that he was clear to fly the third orbit, but warned him once again to manually conserve fuel, as the level was still dropping at an alarming rate. Carpenter contacted the Indian Ocean tracking ship 45 minutes later and reported that he still had 45 per cent of the fuel for his automatic control system and 42 per cent of the fuel for his manual system.

Four hours and 22 minutes after lift-off, now on his third orbit, Carpenter was instructed to prepare to fire his retrorockets and to switch

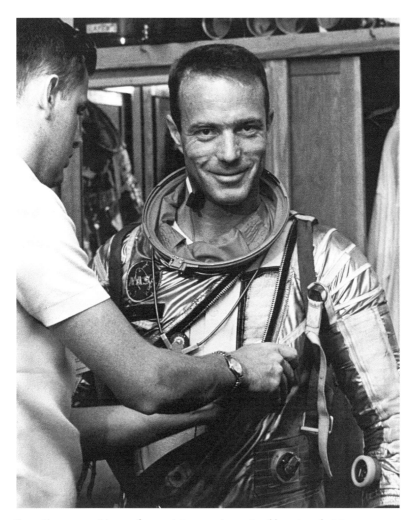

Scott Carpenter suiting up for a training exercise, assisted by suit technician
Al Rochford.

from manual to automatic mode. Twelve minutes later, the retrorockets
were fired and the re-entry phase began, although a problem with the
attitude control system meant that this happened five seconds later than
the designated time, and an overshoot on landing was inevitable. As he
began his descent into the atmosphere, Carpenter continued to report
difficulty with the attitude control system and was monitoring his fuel
level carefully. Nine minutes later, all communication was lost as a layer
of superheated ionized air built up around the spacecraft.[11]

At 1:35 p.m. EDT, Mercury Control reported that the spacecraft might
have travelled about 400 km (250 mi.) from the planned splashdown

site in the Atlantic, saying, 'We expect to re-establish contact with the spacecraft momentarily.' Another seventeen minutes passed in radio silence. Two o'clock came and went, still without any word.

In a live television broadcast to the nation, veteran CBS reporter Walter Cronkite sat stony-faced on camera, giving a running commentary on the efforts to locate *Aurora 7* in the ocean. At one point his concern was evident as he told viewers, 'While thousands watch and pray, certainly here at Cape Canaveral, the silence is almost intolerable.' Then, in a voice choked with emotion, he said, 'We may have lost an astronaut.'[12]

From the moment Carpenter had been expected to splash down, every available resource was thrown into the search. Aircraft scanned the ocean, while the guided-missile destroyer USS *Farragut* raced to the area at top speed. Then, the best possible news came through: a Navy P2V Neptune bomber had received a radio beacon signal, giving the spacecraft's location. Shortly after, Carpenter was spotted in a small life raft he had inflated after crawling out through the top of *Aurora 7*, waving his arms in greeting as aircraft flew overhead. Three paramedics were then dropped from an SC-54 rescue plane, checking Carpenter's condition while keeping him company. They also attached a flotation collar around the spacecraft to keep it upright and afloat.

A jet helicopter then arrived on the scene, having been dispatched from the aircraft carrier USS *Intrepid*, and hoisted the astronaut aboard. A total of 2 hours and 49 minutes had elapsed since he splashed down at 1:41 p.m. The helicopter then transported Carpenter to the waiting carrier, and the destroyer USS *Pierce* was ordered to retrieve *Aurora 7* from the ocean. Carpenter would later be transported to Grand Turk Island for a full medical check-up and to record his post-mission report while every detail was still fresh in his mind.

Despite the multiple problems that Carpenter had to deal with during his flight, he had mostly overcome these and guided his Mercury spacecraft to a safe, if off-target, landing in the Atlantic Ocean. The MA-7 flight also brought the nation an incremental step closer to Kennedy's pledge of a manned lunar landing by the end of the decade.

### Space Twins

At 11:24 a.m. Moscow time on 11 August 1962, a Soviet-built R-7 rocket carrying the Vostok-3 spacecraft thundered into the sky, blazing a trail into the heavens from the remote Baikonur launch complex, carrying 32-year-old cosmonaut Andrian Nikolayev into an elliptical orbit above Earth, sweeping around the planet every 88.5 minutes. He became the seventh person, and the third Soviet cosmonaut, to fly into space. Nikolayev would not remain alone in space for long: less than 24 hours later, a second cosmonaut – Pavel Popovich, 31 – would be launched into orbit at 11:02 a.m. from the same Baikonur launch pad, aboard the companion Vostok-4 spacecraft.

Having achieved the desired orbit, Popovich contacted Nikolayev, who had already completed fifteen orbits, and the two cosmonauts congratulated each other on achieving the first group flight in space. Although the two Vostok craft would never come any closer than 5 km (3 mi.) before inexorably drifting apart, the world nevertheless stood in awe at this tremendous feat of unmatched precision, at a time when the United States had only just begun launching solo astronauts into orbit.[13]

News of this latest Soviet space feat dropped like a bombshell over the United States, where NASA was preparing to send Mercury astronaut Walter Schirra on a six-orbit mission. President Kennedy begrudgingly called the dual flight 'an exceptional technical feat' and saluted the courage of the two cosmonauts. A NASA spokesman termed it a 'tremendous accomplishment', then glumly added, 'It probably will take a miracle for us to beat them to the moon. They've got a great head start – and their great booster power may keep them there.'[14]

On Nikolayev's third day in space, he became the Space Age's first 'million-miler', having completed 49 orbits and travelling around 2 million km (1.24 million mi.) through space. At the same time, Popovich had completed thirteen circuits of Earth and travelled 1.4 million km (870,000 mi.).

Three days after Popovich had been launched into orbit, the Soviet space spectacular came to a successful conclusion when both cosmonauts fired their retrorockets six minutes apart, re-entered the atmosphere, ejected from their spacecraft and landed safely under parachutes south of the town of Karaganda in Kazakhstan. Nikolayev was the first to touch down. The official Russian news agency TASS announced the

Soviet space 'twins' Andrian Nikolayev (front) and Pavel Popovich.

safe return of the two cosmonauts and said they were 'feeling fine'. Although TASS initially reported that both men had landed inside their Vostok spacecrafts, it was later revealed that they had indeed ejected and landed a short distance from their respective charred vehicles.[15]

## A Textbook Mission

Following a near-perfect countdown, the bell-shaped *Sigma 7* Mercury spacecraft (MA-8) lifted off from Cape Canaveral at 8:15 a.m. local time on the morning of 3 October 1962. On board was Walter 'Wally' Schirra, becoming the seventh person to go into orbit, and the third American. He was beginning a space flight scheduled to last six orbits – twice that of the preceding flights undertaken by fellow astronauts John Glenn and Scott Carpenter, and his spacecraft had been modified accordingly.

The primary aim of Schirra's mission was to check the spacecraft's performance and suitability for extended space flights, rather than collecting scientific data. Given this, Schirra would set about conserving electrical power, coolant water and the hydrogen peroxide fuel used in the vehicle's attitude control jets. In order to conserve fuel, he would allow the spacecraft to drift without using the controls. As a result, when it came time for retrofire at the end of his flight, Schirra still had 80 per cent of the ship's hydrogen peroxide propellant remaining – nearly twice the amount required to place *Sigma 7* into the correct re-entry attitude.

The only real concern was during the first orbit when Schirra's space suit became uncomfortably hot, and ground controllers had to consider cutting the flight short. By the second orbit, however, he had managed to bring his cooling system under control and experienced no further discomfort. He would spend around 8 hours and 54 minutes in a weightless state and would later declare that he had suffered no ill effects from his flight. In fact, the entire MA-8 flight became an outstanding exercise in technical competence and was carried out with very little trouble.

At 5:07 p.m. EDT, some 8 hours and 45 minutes after first achieving orbit, Schirra fired his retrorockets, and *Sigma 7* began to dip into the atmosphere on its return journey through the searing heat of re-entry.[16] Slowed by parachutes, the spacecraft came down in the central Pacific (the first ever space explorer to make a Pacific splashdown). Schirra had

MA-8 astronaut Wally Schirra.

guided *Sigma 7* to a landing less than 6.4 km (4 mi.) from the waiting aircraft carrier USS *Kearsarge*, situated around 530 km (330 mi.) northeast of Midway Island. At Schirra's specific request, he would remain in the spacecraft as it was lifted out of the sea and carried across to the *Kearsarge*, where – true Navy man that he was – he wryly requested the permission of the ship's captain to come aboard. Once this fine naval formality had been granted, he blew the hatch and stepped onto the deck at 6:16 p.m.

NASA would later describe Schirra's flight aboard *Sigma 7* as highly successful and said that as a result, the next Mercury flight would attempt a full-day, seventeen-orbit mission (later raised to 22 orbits). A delighted Schirra would also describe his mission as a 'textbook flight', which became the phrase most closely associated with NASA's longest and most productive space flight to that time.[17]

## The Last Mercury Mission

On 15 May 1963, Air Force Major L. Gordon Cooper Jr piloted the final mission in the Mercury series, and the only crewed American space flight launched that year. He would orbit Earth 22 times in a much-modified spacecraft he had named *Faith 7*. It would prove to be the longest American manned space mission to that time, although it had already been exceeded in the earlier flights of cosmonauts Nikolayev and Popovich in their respective Vostok-3 and 4 spacecraft.

The day before his space flight began, Cooper had been strapped inside his Mercury spacecraft for nearly six hours, waiting to be launched. Delays had occurred when the diesel engine on the twelve-storey service tower that stood up against the Atlas rocket failed to ignite. Then came news that a radar breakdown would cause a 24-hour delay in the MA-9 mission. The following morning, tension was high around the Cape as the countdown was resumed and Cooper was once again inserted into the cramped confines of *Faith 7*. He was to remain there for several hours and was likely still exhausted from the tedium of waiting for a launch the previous day. Soon after he had completed his checks, medical staff within Mercury Control noticed a change in the astronaut's bio-readings. Bored by the lack of activity, and with holds being called in the countdown, Gordon Cooper had dozed off.

As launch time grew closer, an amused CapCom Wally Schirra woke the slumbering astronaut. 'Hate to disturb you, old buddy,' he said, 'but we have a launch to do here!' Cooper was instantly awake and alert. 'Let's do it!' he responded.

At 9:04 a.m. EDT, Atlas 130-D lifted off from Launch Complex 14 at the beginning of the planned 22-orbit mission. Five minutes after separating from the 29-metre (95 ft) Atlas booster, *Faith 7* slipped into orbit. The spacecraft would begin circling the planet once every 93 minutes. Throughout his flight, Cooper conducted a variety of tests

Gordon Cooper would fly the final Mercury mission, MA-9.

and experiments and took numerous still and moving images of Earth; manoeuvred *Faith 7* as needed to conduct observations and experiments; and communicated with tracking stations and a number of countries as he passed overhead. He would eat pre-prepared meals and also enter into a programmed sleep period of around 7.5 hours.

Although the flight of *Faith 7* would proceed essentially to plan, the final few hours were filled with high drama after Cooper lost control of vital electrical equipment. As a result, he had to manually manoeuvre the spacecraft into the correct attitude before firing the retrorockets and, while remaining in radio communication, was talked through the re-entry procedures by fellow Mercury astronaut John Glenn. Despite all his life-threatening problems, Cooper remained calm and courageous, bringing his crippled spacecraft through the fiery heat of re-entry to a pinpoint splashdown, remarkably just 5 km (3 mi.) from the main recovery ship, the aircraft carrier uss *Kearsarge*. Frogmen were then dropped from hovering helicopters and they attached a flotation collar to keep *Faith 7* upright in the choppy waters. The Mercury spacecraft, with Cooper still inside, was then hooked up to a crane and winched aboard the *Kearsarge*. Once *Faith 7* had been stabilized and secured, Cooper was given permission to blow his hatch. Apart from some initial dizziness and a severe thirst, he was found to be in excellent condition after his marathon flight.

NASA would later issue a report on its findings into the problems that had plagued *Faith 7*. The primary cause of the failure of the automatic controls was identified as moisture, most likely from Cooper's perspiration, which had corroded a connection on the small electronic box called an Amp Cal (amplifier calibration). The Amp Cal was designed to change electrical signals received from various sensors, such as gyroscopes and infra-red horizon scanners, into firing commands that would ignite small jets in the spacecraft's automatic system. Furthermore, an electrical short circuit in the Amp Cal solder connection – probably due to faulty insulation – had resulted in the loss of other automatic re-entry controls.[18]

Writing in his post-mission report, Cooper observed:

Now that Mercury is over and we stand at the threshold of more ambitious programs, the lessons each of us have learned will be constant tools with which to accept and accommodate new developments. Mercury has been only a beginning for

## U.S. Mercury–Atlas Orbital Missions

| Flight | Pilot | Launch | Landing | Orbits |
|---|---|---|---|---|
| MA-6 | John H. Glenn Jr | 20 February 1962 | 20 February 1962 | 3 |
| MA-7 | M. Scott Carpenter | 24 May 1962 | 24 May 1962 | 3 |
| MA-8 | Walter Schirra Jr | 3 October 1962 | 3 October 1962 | 6 |
| MA-9 | L. Gordon Cooper Jr | 15 May 1963 | 16 May 1963 | 22 |

The MA designation was an abbreviation of Mercury–Atlas, combining the programme name with the Atlas booster. All previous MA flights, numbers 1–5, had been unmanned test missions.

the seven of us. The job at hand is to work to meet our new challenge in space with the same enthusiasm that everyone exhibited throughout this program.[19]

To this day, owing to his Mercury mission, Gordon Cooper remains the last American astronaut to have been launched solo into orbit.

### First Lady of Space

In late 1961, given his role as director of cosmonaut training, General Nikolai Kamanin was ready to recruit a further contingent of pilots for future space missions, but he was also aware of the potential and significant propaganda involved in beating the Americans in the Space Race and displaying superior Soviet technology and ideals to the world. He had read that a group of women pilots (later known as the 'Mercury 13') were undergoing extensive but non-NASA-authorized astronaut training in the United States, and he saw in this a golden opportunity. 'We cannot allow that the first woman in space will be American,' he later wrote in his diary. 'This would be an insult to the patriotic feelings of Soviet women.'[20] He therefore submitted his request for the recruitment of around five women in the next Soviet intake, and this was approved by the Central Committee of the Communist Party. For his part, Khrushchev was delighted with this news, knowing a successful flight by a woman cosmonaut would provide yet another opportunity for the Soviet Union to record another impressive and attention-grabbing space 'first', well ahead of his American rivals, as well as demonstrating the Soviet commitment to women working alongside their male counterparts.

Kamanin knew that the candidates would not need to possess an advanced level of aviation proficiency. The Vostok spacecraft and its systems were of a relatively simple design, and once it had achieved orbit, the majority of functions would be controlled from the ground, which minimized the training time required. During the landing phase, the cosmonaut would be automatically ejected from their Vostok capsule well above the ground, requiring a proficiency in parachuting techniques. He therefore sent out a covert team of recruitment agents to seek out any suitable women with an aerobatic or sport flying background, or those with advanced parachuting qualifications. Eventually, four hundred names crossed his desk, but he quickly ruled out those who did not meet his preferred guidelines. Among those surviving his series of culls was a young textile mill worker named Valentina Tereshkova, who had taken private parachuting lessons.

By 3 April 1962 five suitably qualified women had been selected and approved: Tatyana Kuznetsova, age twenty, a qualified parachutist with a number of world records to her credit; Valentina Ponomaryova, 28, a graduate of the Moscow Aviation Institute, where she had learned to fly and become a parachutist; Irina Solovyova, 24, who had set a number of world records while completing 2,200 parachute jumps; Valentina Tereshkova, 24, a textile mill worker and amateur parachutist who had completed over one hundred jumps; and Zhanna Yorkina,

Vostok-6 pilot cosmonaut Valentina Tereshkova, the first woman to fly into space.

Vostok-5 pilot cosmonaut Valery Bykovsky.

22, also an amateur parachutist. As an early part of their training, all five women received basic cockpit familiarization in order to qualify as flight-trained passengers in turboprop aircraft and MiG-15 dual-seat trainers. They would be permitted to take the controls in order to get a feel for flying, but they were never allowed to fly solo. Having gained this extremely basic flight experience, each of the women subsequently received the commissioned rank of junior lieutenant in the Soviet Air Force.

The flight the women were now training and competing for was designated Vostok-6, while two cosmonauts, Valery Bykovsky and Boris Volynov, were the two prime candidates for the Vostok-5 mission, which would eventually fall to Bykovsky. After much consideration, Kamanin selected Valentina Tereshkova for the Vostok-6 mission. She had scored highly in all her tests, and Kamanin later said of her: 'We must send Tereshkova into space first, and her double [backup] will be Solovyova. Tereshkova . . . is Gagarin in a skirt!'[21] Khrushchev was delighted with Tereshkova's selection, agreeing she had an outstanding profile in terms of propaganda: single, attractive, personable and hard-working, as well as being the daughter of a collective farm worker killed in 1940 during the Winter War between the Soviet Union and Finland.

At 2:59 p.m. Moscow time on 14 July 1963, 28-year-old Lieutenant Colonel Valery Bykovsky became the fifth Soviet cosmonaut to be launched into orbit. Even as the TASS news agency announced this latest space feat, there was informed speculation circulating in the West that his Vostok-5 flight might include a 'cosmic rendezvous' with the first spacewoman. Once Bykovsky was safely in orbit, plans went ahead to launch Vostok-6 two days later.

All of the speculation came to an end with the TASS announcement: 'On 16 June 1963, at 12:30 p.m. Moscow time, a spaceship, Vostok-6, was launched into orbit . . . piloted, for the first time in history, by a woman, citizen of the Soviet Union, Communist Comrade Valentina Vladimirovna Tereshkova.' It added that her call sign would be 'Chaika' (Seagull). As expected, the launch of the first spacewoman created headlines around the world.

During Tereshkova's first orbit, her Vostok spacecraft and Bykovsky's would come within 5 km (3 mi.) of each other. Although highly desirable for propaganda purposes, it was a difficult thing to plan for or predict. Nevertheless, this celestial accomplishment was lauded, although it would be the closest the two spacecraft came to each other during the entire dual mission. Tereshkova would later admit – albeit grudgingly – that as her flight progressed she became physically ill and drained of energy. This impacted on several of her assigned tasks, which included practising a manual reorientation of her spacecraft. It was apparent to those on the ground, and especially a furious Nikolai Kamanin, who later noted in his diary that Tereshkova was not only ill and suffering from headaches but seriously out of her depth with regard to the technical aspects of her mission. Fortunately, her work programme was quite light and with few practical responsibilities, she was able to sleep off some of her illness.

Meanwhile, a decision had been made on the third day of Bykovsky's mission to bring the flight of Vostok-5 to an early end. This was because an underperforming upper stage had placed the spacecraft in a lower-than-planned orbit, and the scheduled eight-day flight duration could not be sustained due to the effects of gravity. A decision was finally made to bring both spacecraft down on 19 June.

As programmed, Vostok-6 began its return journey to the ground on the 48th orbit, but again there would be post-flight criticism of Tereshkova for failing to report retrofire or the successful separation of her craft from the service module. At around 6 km (20,000 ft) from the ground, Tereshkova was automatically ejected from her capsule and parachuted to the ground, touching down at 11:20 a.m. Moscow time some 620 km (385 mi.) northeast of the city of Karaganda in Kazakhstan.

## Soviet Vostok Space Missions

| Flight | Pilot | Launched | Landing | Orbits |
|--------|-------|----------|---------|--------|
| Vostok | Yuri Gagarin* | 12 April 1961 | 12 April 1961 | 1 |
| Vostok-2 | Gherman Titov | 6 August 1961 | 7 August 1961 | 18 |
| Vostok-3 | Andrian Nikolayev | 11 August 1962 | 15 August 1962 | 64 |
| Vostok-4 | Pavel Popovich | 12 August 1962 | 15 August 1962 | 48 |
| Vostok-5 | Valery Bykovsky | 14 June 1963 | 19 June 1963 | 82 |
| Vostok-6 | Valentina Tereshkova | 16 June 1963 | 19 June 1963 | 48 |

*Gagarin was twice promoted during his space flight, from lieutenant to major. No number was attached to the first manned Vostok mission.

Some astonished workers on a collective farm had witnessed her descent and that of her unoccupied spacecraft and ran to the site, joined an hour later by the first of the recovery teams.

A single orbit later, the retrofire sequence was initiated for Vostok-5, eventually bringing Bykovsky back through the atmosphere and ending with a similar ejection and separate parachute landings for the cosmonaut and his spacecraft. He touched down without incident some two hours after Tereshkova, 540 km (335 mi.) northwest of Karaganda, about 800 km (500 mi.) from where Vostok-6 had landed.

While Bykovsky's landing was cause for celebration by the Soviet people, it had already been overshadowed by news of Tereshkova's safe return. Feted across her nation and around the world, her colossal popularity led Khrushchev to exploit the romance between Tereshkova and Vostok-3's Andrian Nikolayev, which had developed during their training. Aware of the propaganda value of such a union, Khrushchev pressured the couple into hurrying things along and hastily arranged a high-profile marriage. On 3 November 1963, amid much publicity, they tied the knot in a Moscow registry office. Seven months later, a healthy baby daughter named Yelena Andrianovna was born to the Soviet Union's celebrated stellar couple. The marriage would, however, later end in divorce.[22]

In late November 1963, America was in mourning for their beloved president John F. Kennedy, who had been gunned down by an assassin in Dallas, Texas. In one reaction to the tragedy, newly installed president Lyndon B. Johnson announced that Florida's Cape Canaveral would be renamed Cape Kennedy to honour the late president. The area containing all the launch facilities at Cape Canaveral would also be renamed as the John F. Kennedy Space Center. Ten years later, following a high-profile campaign by local Floridians, the geographical name of Cape Canaveral was reinstated, although the name of the space centre would remain.

# 4

# STEPPING INTO THE VOID

It was a startling announcement, broadcast over Radio Moscow on 12 October 1964 as the latest crewed Soviet spacecraft was completing its first orbit around Earth. Reporting on the event, the Soviet news agency stated that the Voskhod-1 (Sunrise 1) craft, carrying three crew members, was 'bigger and more comfortable' than the Vostok vehicles previously flown by the earlier cosmonauts. The Russian people were once again swept up in collective and emotional waves of surprise, delight and patriotism. Such news was rapidly transforming their nation into the supreme superpower of the Space Race, seemingly doing things that the Americans could not match, and the citizens responded accordingly with scenes of mass jubilation.

This latest feat came a year after Valentina Tereshkova had entered the history books as the first woman in space, and now there was a new variant of the Vostok spacecraft, carrying three men – the largest space crew to that time. The commander of the mission was announced as 37-year-old Colonel Vladimir Komarov. Accompanying him on this latest venture were two civilians who were not part of the military cosmonaut team, and who had received only minimal training for the flight. One was Konstantin Feoktistov, an eminent space engineer who had been a principal designer of the earlier Vostok spacecraft. The second civilian on board was Boris Yegorov, a physician who specialized in aviation medicine.

At the time, the flight was lauded as a great feat of engineering, with Soviet propaganda smugly pointing out that the Americans were still some time away from launching their next generation of spacecraft, called Gemini, and that this could carry only two crew members. The truth is that the flight of Voskhod-1 was the most dangerous space mission yet, and it could easily have ended in catastrophe. It only came about following continual pressure exerted on Sergei Korolev and his design team by a propaganda-hungry Premier Khrushchev, who was demanding ever bigger space spectaculars.

After the six successful Vostok missions, Korolev had been planning for a new genus of Soviet spacecraft called Voskhod to operate as a two-pilot spacecraft, although as yet it lacked the capacity for achieving rendezvous and docking. But this schedule did not satisfy Khrushchev's need for international recognition and prestige. Khrushchev knew the Americans were pressing ahead with their two-man Gemini programme and wanted to keep ahead by demanding of Korolev that the next space flight carry three crew members.

Despite Korolev's frustration with these incessant demands, he finally came up with an astonishingly simple but audacious plan. Post-flight, this new mission would attract worldwide acclaim, much to the delight of Khrushchev, but it would also receive justifiable criticism once the full facts of this reckless, hazardous and ill-conceived mission were finally known.

The Voskhod spacecraft was basically a modified and stripped-down Vostok capsule. In order to cram three crew members on board, there were no ejection seats. Had an explosion or fire taken place on the launch pad there would have been no means of extracting the trio from their craft. In lieu of bulky space suits, the crew wore only light,

The crew of Voskhod-1 with Chief Designer Sergei Korolev. From left: mission commander Vladimir Komarov, Korolev, Konstantin Feoktistov and Dr Boris Yegorov.

steel-coloured woollen suits and blue jackets, with white headphone caps. Without protective space suits and helmets, the crew would perish within seconds if the cabin air were to be suddenly evacuated through a fault while above the atmosphere – as happened several years later, with fatal consequences. They were also carrying knives in case they had to fend for themselves in the forests of Siberia after landing.[1]

During their flight, the three men were jammed shoulder to shoulder, with very little room to move around, and were only able to conduct very minor experiments. As they were seated perpendicular to the position of the discarded ejection seat, they were forced to crane their necks in order to read the instruments, which were still mounted in their original orientation. Thus the description of this spacecraft as 'bigger and more comfortable' was a complete fabrication. In reality, it was little more than a perilous propaganda stunt, aimed at belittling the forthcoming American efforts.

While it was reported that the crew were feeling fine, this was also actually incorrect. Both lightly trained civilian cosmonauts Feoktistov and Yegorov suffered badly from space sickness during the flight, with Yegorov the worst affected. He reported feeling giddy and ill during the second orbit and had no appetite. It seems the illness was at its worst on the fifth orbit, although Yegorov was later reported as waking from a deep sleep saying he felt much better.

After a little over a day in space, the de-orbiting rockets were fired and the spacecraft returned to a safe parachute landing with the three spacemen aboard. Much to the crew's astonishment, they came back home to a new leadership regime, as Nikita Khrushchev had been deposed from power while they were still in orbit. Leadership of the Soviet Union had fallen to a trio of higher officials: first secretary of the Communist Party of the Ukraine, Nikolai Podgorny; Leonid Brezhnev, second secretary of the Central Committee; and Alexei Kosygin, vice-chairman of the Council of Ministers. On the day the Voskhod crew enjoyed a triumphant welcome to Moscow, they were met and congratulated at Vnukovo airport by two of the three new leaders, Brezhnev (now first secretary of the Communist Party of the Sovet Union – the most powerful position in the Kremlin) and Kosygin, now the new Soviet premier.[2]

In a frank admission in 1990, Sergei Korolev's then deputy, Vasily Mishin, revealed his misgivings of the three-man flight during an interview for *Ogonyok* magazine:

Was it risky? Of *course* it was. It was as if there was, sort of, a three-seater craft and, at the same time, there wasn't. In fact it was a circus act, for three people couldn't do any useful work in space. They were cramped just sitting – not to mention that it was dangerous to fly. But in the West, they drew the conclusion that the Soviet Union possessed a multi-seat craft. It would never have entered anyone's mind there that we would send a crew into orbit without the appropriate means of rescue.

It was good that everything turned out all right. But what if it hadn't?[3]

Proving the day-long flight was little more than a government-initiated propaganda stunt, neither Feoktistov nor Yegorov would fly into space again, while Komarov later became the first victim of space flight when he perished in an enforced landing after a solo mission beset by multiple technical malfunctions.

A little over two weeks after the landing of Voskhod-1, the United States lost its first astronaut when 34-year-old Ted Freeman's T-38 jet trainer hit a flock of Canadian snow geese while on a routine training flight and crashed near Houston's Ellington Field. Freeman ejected, but too late and too close to the ground, and died on impact.

### A Near-fatal Spacewalk

Five months later, only a few days before the first crewed Gemini mission was scheduled to launch, Voskhod-2 blasted off from the Baikonur Cosmodrome, carrying two cosmonauts on another space spectacular that would create broad headlines around the world.

As NASA geared up to send its first two-man crew into space in late March 1965, the Soviet spacecraft Voskhod-2 was placed into orbit on 18 March, carrying Colonel Pavel Belyayev along with thirty-year-old Lieutenant Colonel Alexei Leonov as co-pilot. As with all previous Soviet space missions, the launch, which occurred at 10 a.m. Moscow time from the Baikonur Cosmodrome, was not publicly disclosed until the spacecraft had safely achieved orbit. Once this had been acknowledged an hour after lift-off, the TASS news agency announced that 'a powerful Soviet rocket' had placed the Voskhod-2 spacecraft into orbit, while naming the two cosmonauts on board.

The crew of Voskhod-2: Alexei Leonov (left) and Pavel Belyayev.

The spacecraft's orbital apogee was given as 495 km (308 mi.) – the highest of any manned spacecraft to that time – with a perigee of 173 km (107 mi.), and the orbital period was 90.9 minutes.[4] Later, as Voskhod-2 was entering into its second orbit and passing over the USSR, TASS further reported that cosmonaut Alexei Leonov had 'stepped out' of the spacecraft 'wearing a special space suit with an autonomous life-support system'. Leonov, TASS stated, 'moved up to five metres away from the ship, successfully carried out the range of prescribed studies and observations and safely returned to the ship.' While the first part of their report may have been essentially correct, it was not known for several years that Leonov had come perilously close to not surviving history's first 'walk' in space.

Khrushchev was no longer around to goad Korolev into creating ever-more-spectacular space 'firsts', but the pressure to keep going one better than the U.S. was ever-present. As the Voshkod craft's main hatch had not been designed with exterior activity in mind, it couldn't be opened or closed in space, and nor could the main cabin be depressurized or repressurized. The design bureau's initial reaction was that a spacewalk was unachievable, especially as the flight plan itself was being prepared in haste in order to beat the Americans to the first spacewalk. There was no waiting around for the next generation of spacecraft,

Following the Voskhod-2 mission, Alexei Leonov painted details of his spacewalk.

which would have the facility for this function; Voskhod-2 *had* to be adapted to the task.

To accomplish this, the design bureau added a side hatch and developed a thin-walled but resilient rubber airlock. A little over 2.5 m (8 ft) long, it would unfold outside the spacecraft once orbit had been achieved, and then be inflated against the side hatch by means of pressurized air canisters located on the exterior of the craft. After the cosmonaut had completed the spacewalk he would return to the cabin via the airlock, and once the side hatch had been secured, the airlock could be jettisoned. There was still considerable risk involved; once the Voskhod's hatch had been opened, the only materials preventing a fatal explosive decompression within the spacecraft were two thin layers of rubber and some fabric covering in the airlock. It also meant that the cosmonaut performing the spacewalk had to rely on the limited oxygen supply contained in his backpack.

As egress time approached, the space-suited cosmonauts knew they had to place their trust in the hardware. The airlock was inflated by Belyayev, and then the side hatch was opened. At 11:32:54 a.m. Moscow time, Leonov crawled through the side hatch, which was closed behind him. Belyayev then cautiously opened the airlock's outer hatch, exposing Leonov to raw space. One of the spacewalker's first duties was to mount a film camera at the end of the airlock to record the momentous event. This footage would be retrieved at the end of his excursion. Leonov then floated free of the airlock – 'like a cork from a bottle', as he later described the sensation – spinning at the end of a 5.5-metre (17½ ft) tether.[5] One of the functions he had to perform involved photographing the space-walk using a still camera mounted on his chest. But his space suit had begun to inflate, and he found he could not reach the camera's shutter release, located on his thigh. As the spectacular views swept by below, the only sounds he heard intruding into the eerie silence were his rapid breathing and heartbeat.

After eight minutes of floating, Leonov later related, he clearly felt the volume of his space suit had changed. His fingertips no longer reached the ends of his gloves, and his feet were floating in his boots. His space suit had ballooned and stiffened alarmingly, and he was faced with

Before his troubles began, Leonov was filmed making history's first spacewalk.

a terrible dilemma. His next planned action was to re-enter Voskhod-2 through the inflatable airlock, along with the thick umbilical cord connecting him to the spacecraft. It was already a difficult manoeuvre, but he was now concerned that, with his swollen space suit, he might not be able to squeeze in.

Leonov carefully retrieved the movie camera and began to slide back into the airlock feet first, but his bloated suit quickly became jammed at the entrance and he was stuck. 'It was getting hot,' Leonov later recalled. 'I could feel a stream of sweat running between my shoulder blades and over my eyes. My hands became wet, my pulse quickened . . . It was quite a physical effort.' He continued with his frantic efforts to squirm back into the airlock, but it was all in vain – he was trapped, and history's first spacewalk looked as if it would end when his air ran out and he perished. In such a case, Belyayev, inside Voskhod-2, would eventually have to make the horrifying decision to leave his companion's trapped body outside, to be burned up along with the inflatable airlock during re-entry.

Leonov then tried squeezing in head first, but his body had heated up so much that he was in danger of passing out. Perspiration was stinging his eyes and his helmet visor had fogged up, badly obscuring his view. His only option, he knew, was one filled with great risk: to deflate his space suit. 'The only solution was to reduce the pressure in my suit by opening the pressure valve and letting out a little oxygen at a time as I tried to inch inside the airlock,' he recalled in 2004. 'At first I thought of reporting what I planned to do to Mission Control, but I decided against it. I did not want to create nervousness on the ground. And anyway, I was the only one who could bring the situation under control.'[6]

He began to bleed the air pressure from his swollen suit, and once he had reached the point where he felt it was safe to try accessing the airlock again, Leonov had a second attempt, but once again failed. He then bled out even more air – the pressure in his suit was now down to an extremely hazardous level, starving him of oxygen and causing his heart to beat at a dangerously fast rate. Leonov was rapidly becoming incapacitated and knew he had only one more valiant attempt in him before he passed out forever. Summoning the last of his dwindling strength, Leonov was finally able to squeeze his whole body head first into the airlock. But then he had to perform yet another 'almost impossible' manoeuvre.

> I had to curl my body round in order to reach the hatch to close
> the airlock, so that Pasha [Belyayev] could activate the mech-
> anism to equalize pressure between it and the spacecraft . . .
> Once Pasha was sure the hatch was closed and the pressure had
> equalized, he triggered the inner hatch open and I scrambled
> back into the spacecraft, drenched with sweat, my heart racing.[7]

Belyayev then closed the spacecraft hatch and jettisoned the trouble-
some airlock.

As Leonov began recovering from his exertions, he discovered he
had perspired so heavily that the fluid was sloshing around inside his
suit. From the time he left the cabin until his return, a total of 24 minutes had passed, which he subsequently recalled as the longest half-hour
of his life.

The near-fatal spacewalk would not be the last of the pair's problems. When Belyayev began preparations for a manual re-entry on the
following orbit, the automatic guidance system that was meant to align
Voskhod-2 for re-entry using solar orientation failed. They had to carry
out a manual orientation of the craft and ignite the retrorocket engine
at the correct time. It was a difficult task in their bulky suits, and they
were 46 seconds late in firing the re-entry engine. As a result, they came
down well off course and overflew their planned landing site by 386
km (240 mi.), coming down in a snow-covered Siberian forest over 160
km (100 mi.) north of the city of Perm, in deep snow and jammed
between two large trees. They opened the spacecraft's hatch into freezing conditions.

The two men then spent a miserable night huddled up in sub-zero
conditions, trying desperately to stay awake. These efforts included
drinking some vodka – strictly forbidden cargo they had managed to
smuggle aboard before the launch. The recovery team finally reached
them the next day, but as the trees were too thick for a helicopter to land
nearby they had to spend another night in the forest, this time in comfortable tents with their rescuers. The next day they were kitted out with
skis dropped from a helicopter and finally reached the rescue site. It
would be several more days before their spacecraft could be similarly
recovered.

As expected, the flight was lauded worldwide as successful and
trouble-free. Alexei Leonov was celebrated as the first person to walk
in space and was honoured with the first of his two eventual Hero of

## Soviet Voskhod Missions

| Flight | Crew | Launch | Landed | Orbits |
|---|---|---|---|---|
| Voskhod-1 | Col. Vladimir Komarov<br>Dr Boris Yegorov<br>Konstantin Feoktistov | 12 October 1964 | 13 October 1964 | 16 |
| Voskhod-2 | Col. Pavel Belyayev<br>Lt Col. Alexei Leonov | 18 March 1965 | 19 March 1965 | 17 |

the Soviet Union awards. This enthusiasm was not shared by NASA's astronauts, with Tom Stafford later recalling the misery it caused them, coming as it did mere days before the first crewed flight of the Gemini programme:

> I remember Gus [Grissom], John [Young], Wally [Schirra] and I watched in the crew quarters just a few days before launch, hearing that the Russians had upstaged us again – they had just a little clip of him floating around and said everything was fine. It was only later on, as I got to know Alexei, that we found out he nearly got killed on that mission. They had a lot of malfunctions, landed way out in the boondocks in the Ural Mountains. It was a day before they could even get a helicopter in to drop skis to them to get out of there. But, they said at the time, everything was fine![8]

### An All-Woman Crew

In April 1965, Nikolai Kamanin, director of cosmonaut training, set down the designated crews for a number of follow-on Voskhod missions, culminating in Voskhod-7. For the long-duration Voskhod-3 mission, planned to last ten to fifteen days, the nominated main or prime crew consisted of Boris Volynov and Georgi Katys. The next crew complement, Voskhod-4, was to have comprised two female cosmonauts, Valentina Ponomaryova and Irina Solovyova, both of whom had served as backups for the Vostok flight of Valentina Tereshkova. By this time the health of Chief Designer Sergei Korolev was declining, and he was anxious to move on to the lunar landing programme. He openly expressed his opposition to continuing this and other Voskhod missions, regarding the all-female flight as a propaganda stunt wasteful of design time and spacecraft, and demeaning to the engineering and science behind such missions. He expressed his strong objections but, as expected, was overruled by Kamanin, who had developed the concept of this flight.

In addition to flying the mission, early planning had Solovyova carrying out the first spacewalk by a woman. She had even begun the same extravehicular activity (EVA) training that Alexei Leonov had under-taken. The propaganda potential of a successful mission would have been enormous. Interestingly, although two other women from that group of five, Tatyana Kuznetsova and Zhanna Yorkina, had remained in training for a future flight, the original Voskhod-4 backup crew was all-male, comprising Viktor Gorbatko and Yevgeny Khrunov, although later in the development of the flight Kuznetsova and Yorkina were substituted.

History records that all of these later planned Voskhod missions were cancelled when Sergei Korolev passed away on a Moscow operating table in January 1966 and was replaced as head of the Experimental Design Bureau No. 1 (OKB-1) by Vasily Mishin, his first deputy, who was keen to move away from Korolev's pet Voskhod programme and concentrate on the development of the more advanced series of Soyuz spacecraft. The four women remained in training until 1969, when their group was eventually disbanded, and no further women cosmonaut trainees would be selected for more than a decade.[9]

### Gemini Flies

With the successful completion of Project Mercury, NASA was ready to move on to the next level in its manned space flight programme. Those plans were unveiled in Houston on 7 December 1961 by the director of the Manned Spacecraft Center, Robert Gilruth. In his announcement, he stated that the more spacious, next-generation spacecraft would carry two astronauts on orbital missions, assisting in developing cru-cial rendezvous and docking techniques to be used by astronauts on the later lunar landing programme. At that time, no name had been applied to the two-man project, which Gilruth said would serve as an essential bridge between the Mercury and Apollo programmes, so it was still under the temporary designation of Mercury Mark II.

A later NASA bulletin reported that NASA was in negotiations with the McDonnell Aircraft Corporation of St Louis, Missouri, who – as with the Mercury series of spacecraft – had been nominated as prime contractor. Their vehicle would weigh around 1.8 tonnes (2 tons), or close to twice the weight of the Mercury capsule. It would maintain the high-drag shape of the Mercury capsule and boast around 50 per cent more interior room than its predecessor. Plans then called for the spacecraft to be launched by a new booster, the Air Force's Titan II, constructed by the Martin Marietta Corporation. Preliminary costing estimated the programme coming in at around $500 million, which included the manufacture of about twelve spacecraft, together with Atlas–Agena and Titan II vehicles. As the NASA bulletin stated, the programme objectives were that:

> Two-man flights should begin in 1963–64, starting with several unmanned ballistic flights from Cape Canaveral for tests of overall booster-spacecraft compatibility and systems engineering. Several manned orbital flights will follow. Rendezvous fly-bys and actual docking missions will be attempted in final phases of the program.
>
> This program provides the earliest means of experimenting with manned rendezvous techniques. At the same time, the two-man craft will be capable of Earth-orbiting flights of a week or more, thereby providing pilot training for future, long-duration circular and lunar landing flights.
>
> NASA's current seven astronauts will serve as pilots for this program. Additional crew members may be phased in during later stages.[10]

The new programme needed a name, and the winning submission came from Alex Nagy in the Office of Manned Space Flight at NASA Headquarters in Washington, DC. Nagy not only had the distinction of naming the nation's next manned space programme but was awarded a prize of a bottle of fine Scotch whisky. On 3 January 1962, NASA officially announced that the two-man spacecraft would be called Gemini, the Latin word for 'twins' and the astrological name given to the third constellation of the zodiac.[11]

It would be just over two years later, on 8 April 1964, that the first test flight of a Gemini spacecraft took place, from Launch Complex 19 at the Kennedy Space Center. Carried aloft atop a Titan II booster, the flight went according to plan. Altogether, the burned-out second stage of the Titan II and the attached, uninhabited spacecraft orbited Earth

First launch of an unmanned Gemini craft atop a Titan II launch vehicle.

64 times, although the test part of the mission was declared at an end after just three of those orbits. All of the flight's major test objectives had been successfully carried out, principally that of checking the structural integrity of the Gemini vehicle and the modified booster, while also proving that an astronaut could survive inside the spacecraft.

The Gemini programme picked up momentum in 1965 with a test launch of the unmanned Gemini 2 finally taking place on 19 January. Unlike the first Gemini launch, which achieved orbit, this was a

NASA's first Gemini crew: Gus Grissom and John Young.

suborbital flight with a primary goal of testing the spacecraft's heat shield. All the mission goals were achieved, and NASA could now proceed with the first human-tended mission.

Originally, Alan Shepard had been selected to command the first manned Gemini mission, together with co-pilot Tom Stafford, but Shepard had been suffering bouts of nausea and a loss of balance, which was finally diagnosed as Ménière's disease, a severe inner-ear condition. In October 1963 he was medically disqualified from flying, and to his consternation was removed from the mission. NASA then named Gus Grissom – originally assigned to the second Gemini mission – and 'space rookie' John Young to crew the three-orbit mission. Young was the first of NASA's second astronaut group to be assigned to a flight. The prime objectives for the flight were to demonstrate manned orbital flight in the vehicle; evaluate the performance of spacecraft systems and an astronaut's ability to manoeuvre the spacecraft; check the operation of the worldwide tracking network; and evaluate the post-splashdown recovery system.

During the Mercury programme, the astronauts themselves had been allowed to name their spacecraft, but permission to continue this into Project Gemini was revoked by NASA, who preferred to use the mission name and number as the spacecraft's call sign. This unexpected edict greatly annoyed Grissom, and he decided to make his feelings known by giving his Gemini spacecraft an unofficial name. He happened to read that the Broadway musical *The Unsinkable Molly Brown* was nearing the end of its run, so he decided to make light of losing his Mercury spacecraft to the sea on the MR-4 mission. 'I'd been accused of being more than a little sensitive about the loss of my *Liberty Bell 7*,' Grissom told reporters, 'and it struck me that the best way to squelch this idea was to kid [about] it. And from what I knew about our Gemini spacecraft, I felt certain it would indeed be unsinkable. So John and I agreed that we'd christen our baby *Molly Brown*.'[12] Although many NASA officials were amused by Grissom's touch of levity, he was told to come up with something more acceptable. 'Sure,' he said. 'How about the *Titanic*?' Grissom was well known for his obstinate ways, so a compromise was finally reached. The Gemini 3 spacecraft would unofficially be known as the *Molly Brown*, but future crews were warned they would not be given the same latitude.

On 23 March 1965, Grissom and Young were launched into orbit on the Gemini 3 mission. At lift-off, Grissom had both gloved hands gripping the D-ring that would have triggered his and co-pilot Young's ejection seats during the first fifty seconds of the flight. Young displayed more faith in his commander and their craft by keeping his hands tightly in his lap. After staging was completed, *Molly Brown* slipped into an initial elliptical orbit of 228 km (142 mi.) by 163 km (101 mi.). Just before commencing their second orbit, Grissom fired the two forward-thrusting Orbit Attitude and Maneuvering System (OAMS) rockets, in a test of the crew's ability to alter the height of the spacecraft's orbit above Earth. Grissom not only managed to lower and roughly circularize *Molly Brown*'s orbit but, in doing so, became the first space traveller ever to alter the path of a spacecraft. Throughout the flight he would also conduct a number of orbit-changing manoeuvres designed to test the spacecraft's attitude control systems.[13] Meanwhile, Young was kept busy carrying out tests on different types of water-reconstituted foods and juices for future flights. He also irradiated samples of human blood using electrical power supplied by the spacecraft's systems and fertilized some sea urchin eggs in an experiment on the effects of weightlessness.

Towards the end of the flight, Young offered a small surprise to Grissom – something that Wally Schirra had secretly asked Young to hide in a pocket in the leg of his space suit. It was a corned beef on rye bread sandwich from the popular Wolfie's delicatessen near the Cape,

The Gemini 4 crew: Ed White and Jim McDivitt.

which Schirra had refrigerated overnight. Grissom took the sandwich with a big smile and took a bite, but as crumbs began to float everywhere he hastily stowed it in his suit pocket. It was worth a chuckle at the time but would later earn both astronauts a severe reprimand for smuggling unauthorized food on board that could interfere with the results of their specially prepared diet.

In a final orbit-altering manoeuvre prior to preparing for re-entry, Grissom fired the forward thrusters, which brought their lowest point of orbit down to around 80 km (50 mi.). He then fired the four solid-fuelled retrorockets to begin *Molly Brown*'s return to Earth. *Molly Brown* splashed down in the Atlantic with a welcome jolt – and proved as unsinkable as her namesake. Grissom released the main parachute and decided they would remain on board as planned and be lifted over to the waiting carrier by helicopter. To his surprise, he was told USS *Intrepid* was around 93 km (58 mi.) distant. Navy frogmen arrived soon after and attached a flotation collar to the spacecraft, by which time it had grown uncomfortably warm inside *Molly Brown*, and the two astronauts were becoming increasingly nauseous as the spacecraft pitched and rolled in the ocean swells. 'That was no boat,' Young later declared.

The maiden piloted flight in the Gemini series had been completed with very few problems, and NASA was already planning for the next mission, Gemini 4. Post-flight, John Young would have the remains of the offending sandwich encased in plastic, placing it on display on his office desk alongside his official letter of reprimand.

### An American Walks in Space

On 27 July 1964, NASA announced that the crew for the Gemini 4 mission the following June would be composed of two first-time flyers: Captain James McDivitt as command pilot and Captain Edward 'Ed' White II as pilot. Early mission plans called for McDivitt to open the spacecraft hatch and poke his head out while they orbited Earth. Although a full spacewalk had been planned for a later Gemini mission, the largely unexpected EVA by Alexei Leonov in March 1965 caused these plans to be accelerated. Both men began full EVA training, but ultimately it was decided that White would attempt America's first spacewalk. The mission would also become the first flight to be controlled from the new Mission Control Center at NASA's Manned Spacecraft Center in Houston.

The second crewed mission in the Gemini series lifted off from Launch Pad 19 at Florida's Kennedy Space Center on 3 June 1965. The two-stage Titan II booster rocket successfully carried the Gemini spacecraft through to orbit, following which McDivitt turned the ship around with short bursts of the thrust engines, located in the sides of the adapter module. He then attempted to rendezvous with the orbiting, fuel-depleted second stage of the Titan rocket but aborted this when he found he was using far more fuel than he had estimated, and Mission Control agreed.[14]

White's EVA was scheduled for the second orbit, and he began preparing himself for the attempt. However, the astronauts had been so busy on the failed rendezvous that it was agreed not to rush things and to delay the EVA until the third orbit. Having achieved this, the two astronauts sealed their helmets and reduced the pressure inside the spacecraft to zero. White then opened the hatch on his side and stood up on his couch. Four minutes later, having established everything was in order, he grasped the edge of the hatch and hauled himself out of the spacecraft, 217 km (135 mi.) above Earth. He was attached to the Gemini craft by means of a 7.6-metre (25 ft) umbilical tether covered in insulating gold tape. This lifeline would also provide him with his oxygen supply and lines of communication. White then cautiously used bursts on a small oxygen-jet gun to propel him to the front of the spacecraft, where McDivitt began photographing the historic EVA, meant to last up to twelve minutes.

As White floated around their craft, he reported that he was suffering no symptoms of disorientation. Quite the opposite – he was enjoying himself immensely. In fact, he was so enthralled by the experience that he overstayed the time allowed and made light of McDivitt's warnings. Several times, CapCom Gus Grissom told McDivitt to get White back inside the spacecraft, as the ground station at Bermuda was coming into range and Houston would soon lose communication with Gemini 4. Reluctantly, White moved back inside, but then encountered problems fastening his hatch, which meant that the cabin could not be repressurized. It was eventually resolved, although, out of caution, a second hatch opening a little later to dispose of some EVA equipment was cancelled, which left the astronauts with less room in their cabin than expected.

In all, Ed White's EVA had lasted 23 minutes – far too short for the astronaut. He would later state that the spacewalk was easily the most

Ed White photographed during his historic Gemini 4 EVA.

comfortable part of the entire mission and that the order to end it and get back into the spacecraft was the 'saddest moment' of his life.[15]

Splashdown of the Gemini 4 mission in the Atlantic Ocean took place around 68 km (42 mi.) from the targeted area due to a failure in the craft's landing system computer. Soon after, Navy parachutists dropped from a waiting helicopter and attached a flotation collar around the spacecraft, following which the hatch could be opened. The two grinning astronauts were then hauled up into a second helicopter and transported across to the recovery carrier, USS *Wasp*, for medical checks, debriefing and a well-earned rest.

By this time, Gordon Cooper and space 'rookie' Charles 'Pete' Conrad were well into preparations for the eight-day Gemini 5 mission, which was not only expected to better the Soviet space flight endurance record by three days but intended to test new fuel cells and to replicate the time it would take for Apollo astronauts to fly to the Moon and

Pete Conrad and Gordon Cooper: the crew of Gemini 5.

back, allowing NASA physicians to evaluate the effects on a crew during a lengthy period of weightlessness. Their spacecraft would be equipped with its own radar system, which the astronauts would use during a rendezvous attempt with a small electronic object they would release, allowing it to drift away. The football-sized object, officially the Radar Evaluation Pod (or REP, but humorously called 'Little Rascal' by the crew), was equipped with a radio transponder and receiving beacon as well as a flashing light. If successful, the tracking of and catch-up with another orbiting object would prove crucial in NASA's plans for a manned lunar landing by the end of the decade, in which rendezvous and docking techniques were essential.[16]

Because of postponements, the launch date eventually slipped to 21 August 1965, but on that day there was a perfect launch and insertion into orbit. Just over two hours into their flight, Cooper and Conrad released the REP from its position in the rear of the spacecraft's adapter section. After receiving a solid radar fix on the pod they allowed it to drift away for a later attempt at orbital rendezvous. Five hours and 34 minutes into their flight, Conrad told Mission Control that the pod was some 600 m (approx. 2,000 ft) from them and still visible.

By this time, the crew had also reported a serious problem with a fuel cell, which could have impacted adversely on the production of electricity to the spacecraft. They were advised to power down while everyone worked on the problem. The pod chase was eventually called off, as its battery had run down while the crew's problems continued. To everyone's relief, the fuel cell problem was finally resolved, full power was restored, and the flight was allowed to continue. An alternative plan was devised for the radar system, according to which the astronauts would rendezvous with a phantom target in an imaginary orbit. Later in the mission they were able to carry out this manoeuvre with little difficulty, arriving at the rendezvous point with reasonable precision.

As the days passed and space endurance records began to tumble, Cooper and Conrad began to feel the lesser consequences of flying an eight-day mission. Cooper began to show signs of irritability, criticizing the planning of some of the experiments set for them to conduct and saying they were not being given enough time. As Conrad would also relate post-flight:

> The romance ended fairly quickly. We were really confined in an extremely small space. My knees began to bother me. It felt as though my knee sockets had gone dry. I hurt and I didn't want to stay in there. If they had told me I had to stay up longer than the eight days, I believe I would have gone bananas. My body ached, and my mind was not active enough. I was the rookie and Gordo was the pro. We had trained together for a year and there weren't any stories left to exchange. We had some systems failures, and this precluded us from doing some of the tasks we were supposed to do, so all we could do is sit. You can't go to sleep; you just don't get tired. Your body is uncomfortable, you don't do any work, and zero G makes you lethargic.[17]

Towards the end of their flight, Cooper and Conrad had eclipsed the Soviet Union's cumulative cosmonaut time in space of 507 hours and 16 minutes, and beaten the single flight record of Valery Bykovsky on Vostok-5, surpassing the 119 hours and 6 minutes the cosmonaut had recorded on his flight. On 29 August 1965, having completed 120 orbits of the planet in 180 hours and 56 minutes, the Gemini 5 mission came to an end with a successful splashdown in the western Atlantic Ocean, and medical examinations showed the two astronauts were in excellent condition following their marathon journey through space.

### Problems, and How to Rendezvous in Space

Over its first three crewed flights, Project Gemini had been scoring an impressive number of achievements, and Gemini 6, to be crewed by Wally Schirra and Tom Stafford, was planned to provide yet another crucial step towards the Apollo programme. They were scheduled to rendezvous and dock with a modified Agena rocket upper stage, known as the Agena Target Vehicle (ATV) launched just 90 minutes ahead of the crewed spacecraft.

The Agena was an uncrewed spacecraft fitted with a docking collar at one end, and was used by NASA to train crews in orbital rendezvous techniques. It was a 7.92-metre-long (26 ft) cylinder with a diameter of 1.52 m (5 ft). Launched atop an Atlas booster, as it neared orbital status a two-piece protective shroud would be jettisoned, following which the Agena was explosively separated from the carrier Atlas rocket before entering a low circular orbit.

On 25 October 1965, Schirra and Stafford were sealed inside their Gemini craft in preparation for their launch, eager to hear news of the successful launch of the nearby Atlas rocket carrying their Agena target. That lift-off came right on schedule, but six minutes into the ascent, when the Agena's main engine fired to separate it from the Atlas, there was a huge explosion, showering debris into the Atlantic. On Launch Pad 19, Schirra and Stafford were informed of the loss of the Agena and that their own launch had to be postponed.

Several contingency plans were considered, until McDonnell's Gemini spacecraft chief Walter Burke suggested to NASA that, while it would take some time to replace the Agena, there was another orbital mission undergoing advanced preparation – Gemini 7. He suggested this long-duration flight, crewed by Frank Borman and James Lovell, be launched as scheduled, followed soon after by a second launch, that of Gemini 6. The two spacecraft would rendezvous in orbit – proving that capability – but not dock. Initially reluctant, NASA eventually came around to Burke's way of thinking, undoubtedly fuelled by the enthusiasm of the four crew members.

The Gemini 6 crew: Tom Stafford and Wally Schirra.

The crew of Gemini 7: Jim Lovell and Frank Borman.

Moments of tension: the launch abort of Gemini 6A.

Following additional mission-specific training, the Gemini 7 crew lifted off from the Kennedy Space Center on 4 December 1965, achieving orbit eight minutes later. Borman and Lovell then settled in, carrying out their duties while preparations continued on the ground for the launch of Gemini 6A (as it was now renamed) eight days into their flight.

On 12 December, after a mammoth effort, Schirra and Stafford were once again seated aboard their spacecraft as the countdown clock wound down to zero, just as Gemini 7 was completing its 117th orbit. As the countdown reached the minus-three-second point, the Titan's engines fired. Then, 1.6 seconds before lift-off, it all came to an abrupt end as the engines suddenly shut down.

Following procedure, both astronauts had their hands on an ejection handle positioned between their legs, and either man could have initiated an emergency ejection of both astronauts. The hatches would have slammed open and the crew been explosively blasted out of their spacecraft, parachuting to safety well away from the launch pad. The sudden interjection 'Shutdown Gemini 6!' was relayed to the crew, who now faced a massive dilemma, but neither man flinched. They were bravely gambling with their very lives, not knowing whether the mighty charge of fully pressurized propellant beneath them was about to explode. They waited as the seconds ticked by, watching their instruments and holding tight until the pressure dropped to a safe level and they could be extracted from the spacecraft. It was a gutsy decision that could not only have cost them their lives but shut down the Gemini programme for an indeterminate length of time.

Meanwhile, Gemini 7 was passing directly overhead. The crew would have had a good view of the launch but noted that it had not taken place as scheduled. Once they were told what had happened and that the crew was safe, Borman radioed: 'We saw it ignite. We saw it shut down.'[18]

Technicians launched an immediate investigation. The cause was quickly identified as a small electrical plug, worth just a few cents, that had dropped out prematurely from the base of the Titan rocket. This triggered the mechanism that detects problems and shut down the engines. Despite all that had happened, it was only a minor fix, and plans were soon under way for yet another launch attempt.

Just three days later, on 15 December, the launch of Gemini 6A finally took place, and seven hours later the two Gemini spacecraft had achieved the planned rendezvous. The two ships manoeuvred around each other for several orbits, with both crews reporting it was an easy operation. The closest approach was within 30 cm (1 ft) of each other

The Gemini 6A spacecraft as seen by the crew of Gemini 7 during their historic orbital rendezvous.

– so close the astronauts could see each other through their windows. Had the spacecraft been equipped with docking devices, they could easily have linked up.

After just 26 hours of space flight and sixteen orbits, Schirra initiated their scheduled re-entry, splashing down northeast of the Turks and Caicos Islands just 18 km (11 mi.) from the planned site, where they and their spacecraft were retrieved by helicopter and transported across to the recovery carrier uss *Wasp*. Two days later, at the end of their marathon fourteen-day space flight, and having completed a record-breaking 206 orbits of the planet, Borman and Lovell also splashed down. They too were recovered by the crew of the *Wasp*, although by that time Schirra and Stafford had been whisked back to the Kennedy Space Center. Although setting a new endurance record, two weeks spent inside the cramped Gemini 7 spacecraft constituted a different kind of endurance. As Lovell reported after the flight, it was like spending two weeks stuck in a men's room.

### Space Emergency

Plans that had once been in place for Gemini 6 to rendezvous and dock with an ATV now fell to the crew of Gemini 8, Neil Armstrong and David Scott. Scott would also undertake America's second tethered spacewalk, remaining outside for at least 95 minutes, while completing a single orbit of Earth. Utilizing a handlebar-shaped gas-propelled gun, he would then make his way to the rear of the white adapter section, where he would attach himself to a backpack – essentially a long-life oxygen tank – and exchange his 8-metre (26 ft) tether for one three times longer.

When the Atlas–Agena lifted off and cleared the launch tower from Launch Pad 14 on 16 March 1966, Armstrong and Scott were watching from inside their Gemini spacecraft, sitting on the nearby Launch Pad 19. Then, 101 minutes later, they were also launched into orbit atop their Titan II rocket, ready for the space chase to follow.

Scott first sighted the Agena over the Indian Ocean near the northern tip of Indonesia, around 122 km (76 mi.) away, and they rapidly closed the distance. Four hours after they had lifted off, Armstrong and Scott caught up to the Agena. After circling the target vehicle to ensure it had not suffered any damage during launch, Armstrong eased their Gemini craft to a point where he could slide its nose into the Agena's docking collar. 'It was a real smoothie,' Armstrong said as – for the first time in history – two spacecraft were linked together in orbit.

Soon after, however, a rogue thruster on the Gemini craft suddenly began firing, rotating the conjoined vehicles. Armstrong immediately fired their thrusters to correct the spinning, but soon after the problem reoccurred. Believing the fault was with the Agena, Armstrong shut down its attitude control system, but this did not resolve the situation. Things got even worse when the two spacecraft began rotating in two axes. They decided to undock and move away from the Agena, but this only made things worse. Once released, the Gemini craft began spinning even faster, rotating at one revolution per second. 'We were in an uncontrollable tumble in space,' Scott later wrote. 'And it was about to get much worse.'[19]

With the spin rate accelerating and their spacecraft now toppling end over end, the astronauts were in serious danger of blacking out. Despite the uncontrolled spinning, they continued to communicate their actions to Mission Control. They finally managed to shut down the manoeuvring rockets – including the faulty unit – and fired the Re-entry Control System (RCS) thrusters to control the spin. In doing so, they consumed around three-quarters of their re-entry manoeuvring fuel, which under mission guidelines meant that the mission had to be

The crew of Gemini 8 (both later moonwalkers), Neil Armstrong and Dave Scott.

abandoned. Mission Control thus told them to end the mission and re-enter as soon as possible. Armstrong and Scott then went through their lengthy re-entry procedures before firing the retrorockets and discarding the spacecraft's adapter section as they passed over Africa for the seventh time.

Gemini 8 splashed down in the Pacific Ocean, 800 km (500 mi.) southeast of Okinawa, Japan. Having landed so far away from the recovery fleet, the astronauts had to spend three hours in heaving seas while waiting for a Navy destroyer to arrive and pick them up. Eventually USS *Leonard F. Mason* arrived on the scene and soon had the exhausted, seasick astronauts and their spacecraft safely on board.

The most probable cause of the erratic firing was traced to an electrical short-circuit that kept the thruster firing for three seconds, then shutting down before firing again, even when switched off. On later missions each thruster would have an isolated circuit that could be switched off. Armstrong and Scott were subsequently praised for their calmness and expertise in recovering from a potentially fatal crisis. Both men would later command Apollo missions and walk on the Moon.

### A Backup Crew Takes Over

It was a clear winter's morning in Houston on 28 February 1966 when the Gemini 9 crew of Elliot See and Charles Bassett took off from Ellington Air Force Base, near the Manned Spacecraft Center, in one of NASA's T-38 jet aircraft, together with their backup crew of Thomas Stafford and Eugene Cernan in a second T-38. The four astronauts were flying to the McDonnell plant in St Louis, where the Gemini craft assigned to their mission was nearing completion.

With the abandonment of Dave Scott's spacewalk on Gemini 8, Bassett was in line to become NASA's second spacewalker. The two jets took off from Houston in perfect weather conditions, but this would not be replicated in St Louis, where the forecast had grown increasingly bad, with low visibility, rain, fog and a broken ceiling of just 183 m (600 ft). The weather on their arrival was even worse, and while See opted to fly below the clouds to check conditions, Stafford decided to climb out of the threatening clouds and snow flurries. Unfortunately, See descended too low in trying to get a visual impression of the runway. Their T-38 hit the roof of the McDonnell plant, where the Gemini spacecraft was housed, and bounced and crashed into an adjacent car

At front, the original Gemini 9 crew of Elliot See (left) and Charlie Bassett. The backup crew of Tom Stafford and Gene Cernan would take on the GT-9A mission following the loss of the prime crew.

park, where it exploded. Both men had been killed instantly when the aircraft hit the roof of the building.

Once NASA had recovered from the aftermath of this tragic accident, Stafford and Cernan were named as the replacement crew to fly the Gemini 9 mission. To Cernan fell the task of carrying out the EVA, previously assigned to Charles Bassett. This entailed the use of what was called the Astronaut Maneuvering Unit, or AMU, which would be loaded pre-flight within the white adapter section at the back of the spacecraft. The AMU, developed by the U.S. Air Force, was essentially a propulsive backpack equipped with hydrogen peroxide thrusters and two hand controllers. Cernan's job during his EVA was to make his way to this stowage and don the AMU before carrying out a number of simple preassigned tasks to test its capability and suitability for work-related tasks outside an orbiting vehicle.

For mission commander Tom Stafford, history would repeat itself on 17 May 1966 when, once again, an Agena target rocket was lost after take-off, just as he had experienced seven months earlier on the Gemini 6 mission with Wally Schirra. The Atlas carrier rocket blasted off without problem, but five minutes into the flight one of the two outboard booster engines swivelled sharply, pulling the rocket off its planned trajectory. Unfortunately, the failure came only half a second before the boosters were due to be jettisoned. The ATV separated as scheduled but headed off in the wrong direction. Both rockets then tumbled into the Atlantic.

Eventually, Gemini mission director William Schneider made the solemn announcement: 'We have lost our bird. The mission is scrubbed. Gemini will not fly today.'[20] In fact the mission, now renamed Gemini 9A, would remain grounded for more than two weeks as an alternative flight plan was assembled.

With no substitute Agena available, Stafford and Cernan were told they would not be flying the full original mission, although the EVA would remain as planned. They would now be pursuing a less sophisticated target vehicle: an Augmented Target Docking Adapter (ATDA). As the ADTA had no manoeuvring power, the once-scheduled exercise of docking with the Agena and using it to power the Gemini spacecraft – basically acting as an orbiting fuel tanker – was abandoned.

The ADTA was successfully placed into orbit on 1 June 1966. Gemini 9A should have followed soon after but suffered another frustrating delay caused by the failure of a computer system within Mission Control. Although the fault was traced and repaired within hours, the flight had to be postponed until the ADTA was in position for a successful rendezvous. In a further blow, tests had shown that while the ADTA had achieved a perfect rendezvous orbit, a fibreglass aerodynamic shroud at the front of the satellite had failed to fully separate. If this was the situation, the crew would be unable to achieve an orbital docking.

Two days after the ADTA launch, it was third time lucky for Stafford and Cernan. On 3 June, they reached the desired orbit 6 minutes and 20 seconds after launch. Following a four-hour chase, Stafford caught up with the ADTA and confirmed that the shroud was still jammed in place and partially open. 'I've got a weird-looking machine in sight,' he reported to Houston. 'It looks like an angry alligator rolling around out here . . . it's moving all around.'[21]

Plans were discussed for Stafford to try nudging the ADTA, hoping to jar the shroud into separating, and there was also a suggestion that Cernan try to separate the open shroud manually during his spacewalk, but both plans were rejected as too hazardous. After circling the ADTA for a while, Stafford finally put some distance between their spacecraft and his 'angry alligator' as a safety precaution.

Gene Cernan's spacewalk began on 5 June, and when he finally emerged from the Gemini spacecraft on his planned two-hour spacewalk he became only the second American astronaut to conduct an EVA. He was not to know it, but this undertaking would almost cost him his life. The stiff, pressurized space suit greatly restricted his movements as he slowly made his way to the rear adapter section where he was supposed to strap on the waiting AMU. Another problem he encountered was a lack of handholds and physical supports as he headed towards where the unit was stowed. Donning the AMU proved to be such an incredibly difficult task that it sapped him of what remaining strength he had, and to make matters worse his suit's rudimentary cooling system was ineffectual and caused overheating for Cernan. As a consequence, the inside of his faceplate was fogging over, obscuring his view. Exhausted, seriously hot, almost blind, and with no means of wiping his faceplate clean, Cernan was in serious trouble. 'Lord, I was tired,' he later wrote. 'My heart was motoring at about 155 beats per minute . . . I was

Tom Stafford's 'angry alligator'.

sweating like a pig.' He could no longer continue with his EVA and abandoned the effort to don the AMU. Somehow, he managed to claw his way back to the open hatch, where Stafford helped guide him back inside and close the hatch. 'I was as weary as I had ever been in my life,' Cernan said.[22]

Having orbited Earth 47 times in the course of their three-day mission, the two disappointed astronauts splashed down safely on 6 June and were recovered by the crew aboard the aircraft carrier USS *Wasp*.

### The Final Gemini Mission

There would be three more Gemini missions, all successfully completed, albeit with several problems to solve. There would be further dockings with ATVs, using the propellant within them to soar to record heights, the most notable occasion being when the Gemini 11 crew of Pete Conrad and Dick Gordon fired the docked Agena's engine, boosting their orbit to an altitude of close to 1,370 km (850 mi.). Spacewalks were also performed by astronauts Michael Collins, Dick Gordon and Buzz Aldrin.

When the crew of Gemini 12 lifted off from the Kennedy Space Center's Launch Complex 19 on 11 November 1966, Navy Captain James Lovell and Air Force Major Edwin 'Buzz' Aldrin were ready to see out the concluding flight in a programme of astonishing accomplishments. In the space of just twenty months, beginning with Gemini 3 in March 1965, a total of ten two-man missions had left the launch pad, at an astounding average of one every two months.

One of the major tasks designated for the four-day flight was an attempt to solve major problems that had plagued every EVA since Ed White's almost task-free twenty-minute spacewalk back in June 1965. The space agency had come to the realization that performing manual work outside the spacecraft was an exhausting and therefore potentially hazardous operation, most recently having caused a badly fatigued Dick Gordon to shorten his space walk on the previous mission, Gemini 11.

In order to familiarize himself with the conditions he might encounter during the final Gemini mission, Aldrin began practising EVA operations in a specially equipped swimming pool at the Environmental Research Associates facility near Baltimore, Maryland, which was equipped with a full-size mock-up of a Gemini spacecraft's equipment

The Gemini 10 mission carried John Young (left) and Mike Collins to a new record altitude.

section and Agena docking collar. Wearing a modified space suit equipped with weights and floats, he trained in a reasonable approximation of conditions he would confront on his upcoming spacewalk.

An hour and a half before their launch, an Agena rocket had blasted a steady path into orbit to serve as a target for planned rendezvous and docking. Once the Gemini craft had also left the launch pad, the two astronauts began a 103,000-kilometre (64,000 mi.) 'tag game' with the orbiting Agena, finally accomplishing a firm docking 4 hours and 13 minutes after lift-off.

On the second day of their flight, having carried out a docking with their orbiting Agena the previous day, Aldrin opened the spacecraft hatch before undertaking a limited 2-hour, 13-minute EVA in which he stood up through the hatch and performed a number of simple tasks, such as conducting ultraviolet astronomical photography. He also installed a handrail between the open hatch and the docking collar of the Agena to assist him during his full spacewalk the following day.

The next day, as Jim Lovell remained inside the Gemini craft, Aldrin once again opened the hatch and made his way out, spending a record 2 hours and 9 minutes performing tasks that had baffled and exhausted previous spacewalking astronauts. Unlike them, he did not suffer any

Another altitude record was achieved by the crew of Gemini 11, Dick Gordon (left) and Pete Conrad.

excess fatigue, nor did his faceplate fog up, and he did not have to cut short his EVA.

Aldrin proved that, with the right approach to the task, and preparatory underwater training, an EVA could be accomplished with relatively few problems. From a practical point of view, additional handrails and handholds were now in place, along with a waist tether that enabled Aldrin to turn wrenches and retrieve experiment packages without overexerting himself. He also allowed himself more time to complete his tasks and would take regular rest breaks. Aldrin later attributed his successful spacewalk to the many hours he had spent practising in the pool at Baltimore. Eventually, their work done, the two astronauts were able to relax, with little to do but drift, talk and listen to music before the time came to prepare for re-entry.

After 59 orbits, as millions of viewers watched on live TV on 15 November, Gemini 12 splashed down in the western Atlantic in a pin-point landing so accurate that a recovery helicopter was seen hovering above the spacecraft as it made its final parachute descent, and the recovery carrier, USS *Wasp*, was stationed just 5 km (3 mi.) away. Within half an hour, the two beaming astronauts had been deposited on the deck of the *Wasp*.

Thus ended Project Gemini, the most successful crewed space venture to that time. Over those twenty eventful months, almost every concept required to fly astronauts to the Moon had been proven. As NASA would proudly relate following the successful conclusion of Gemini 12:

> The Gemini program consisted of a total of 19 launches, 2 initial uncrewed test missions, 7 target vehicles, and 10 crewed missions, each of which carried two astronauts to Earth orbit. Designed as a bridge between the Mercury and Apollo programs, the Gemini program primarily tested equipment and mission procedures and trained astronauts and ground crews for future Apollo missions. The general objectives of the program included: long duration flights; testing the ability to maneuver a spacecraft and to achieve rendezvous and docking

The crew of the final mission in the Gemini series, Buzz Aldrin (left) and Jim Lovell.

## U.S. Crewed Gemini Missions

| Flight | Crew | Launch | Landed | Orbits |
|---|---|---|---|---|
| GT-3 | Virgil I. 'Gus' Grissom<br>John W. Young | 23 March 1965 | 23 March 1965 | 3 |
| GT-4 | James A. McDivitt<br>Edward H. White II | 3 June 1965 | 7 June 1965 | 62 |
| GT-5 | L. Gordon Cooper Jr<br>Charles 'Pete' Conrad Jr | 21 August 1965 | 29 August 1965 | 120 |
| GT-6A | Walter M. Schirra Jr<br>Thomas P. Stafford | 15 December 1965 | 16 December 1965 | 16 |
| GT-7 | Frank F. Borman II<br>James A. Lovell Jr | 4 December 1965 | 18 December 1965 | 206 |
| GT-8 | Neil A. Armstrong<br>David R. Scott | 16 March 1966 | 17 March 1966 | 6 |
| GT-9A | Thomas P. Stafford*<br>Eugene A. Cernan | 3 June 1966 | 6 June 1966 | 47 |
| GT-10 | John W. Young<br>Michael Collins | 18 July 1966 | 21 July 1966 | 43 |
| GT-11 | Charles 'Pete' Conrad Jr<br>Richard F. Gordon Jr | 12 September 1966 | 15 September 1966 | 44 |
| GT-12 | James A. Lovell Jr<br>Edwin E. 'Buzz' Aldrin Jr | 11 November 1966 | 15 November 1966 | 59 |

*Original GT-9 crew of Elliot M. See Jr and Charles A. Bassett II killed in plane crash

of two vehicles in Earth orbit; training of both flight and ground crews; conducting experiments in space; extravehicular operations (standup sessions and spacewalks); active control of reentry to achieve a precise landing; and onboard orbital navigation.[23]

Project Gemini had provided a tremendously accomplished springboard into NASA's lunar landing plans, and the next crewed mission was planned to fly atop a mighty Saturn rocket in an Apollo spacecraft.

Meanwhile, there was growing speculation that the Soviet Union might have deferred plans for any further involvement in the so-called Space Race to the Moon (although this would later prove to be false). When Gemini 12 brought the two-man programme to a successful conclusion, it had been some eighteen months since the Russians had launched their last crewed mission, Voskhod-2. During that period, NASA had sent eight two-man Gemini flights into orbit – equal to the total number of manned spacecraft launched to that time by the Soviet Union. Amid mounting excitement, Americans were now looking to NASA to achieve President Kennedy's goal of landing astronauts on the Moon by the end of the decade – and also achieving this ahead of the Russians.

# 5

# TRAGEDY ON THE LAUNCH PAD

By the end of 1965, NASA's focus was increasingly turning from the highly successful Gemini series of missions to Project Apollo. With the first crewed orbital flight scheduled before the end of the following year, the space agency's director of flight crew operations, Deke Slayton, decided it was time to select the first Apollo crews.

Slayton's initial choice to command the maiden flight, an Earth-orbiting test of the Apollo spacecraft, had been his fellow Mercury astronaut Alan Shepard. However, Shepard's flying future was in serious doubt after he had contracted the debilitating inner-ear malady Ménière's disease, so Slayton had to look elsewhere. 'Gus Grissom was going to be coming off the backup assignment to GT-6A,' he later revealed in his memoirs, 'and so was a pretty natural choice for commander of the first mission.'[1]

America's first spacewalker, Ed White (Gemini 4), was another name Slayton pencilled in for the first Apollo crew. As no lunar module would be involved in this test flight, Slayton felt he could assign a relatively inexperienced crewman to fill the third seat, and his choice came down to two candidates: Donn Eisele and Roger Chaffee, both Group 3 astronauts who had earlier been paired on tests of the lunar space suit's life support systems. As the choice came down to a question of crew compatibility, Slayton determined that Eisele might prove the better fit. Unofficially, he now had his first crew, although procedurally the names would have to be submitted to NASA headquarters for official approval.

Then, before the provisional crew names had been referred to NASA, fate stepped in to bring about a crucial change. Months earlier, in September 1964, Eisele had dislocated his left shoulder while participating in zero-gravity training aboard a NASA KC-135 aeroplane. The injury had healed, but in January 1966 Eisele dislocated the shoulder a second time, while undertaking some physical exercises. Much to his chagrin, he was removed from flight status, and Slayton replaced him on the first Apollo crew with Roger Chaffee. This is the crew that Slayton would eventually submit to NASA headquarters. Meanwhile, Eisele was given additional time to overcome his shoulder injury and found himself provisionally assigned to the second Apollo mission, along with Wally Schirra and Walt Cunningham. All of these crew changes were confirmed to the author by Eisele's first wife, Harriet Eisele, Walt Cunningham and, prior to his death in 2007, Wally Schirra.[2]

Originally, the first Apollo mission was known as AS-204. The 'A' stood for Apollo and the 'S' for the Saturn IB launch vehicle (which was also the number 2 in the flight designation). The '04' signified that it would be the fourth launch of an Apollo mission. While AS-204 was the official NASA designation for the flight, everyone knew it less informally as Apollo 1.

The crew assigned to the Apollo 1 (AS-204) mission: Ed White II (left), Gus Grissom and Roger Chaffee.

For Gus Grissom, AS-204 would provide the chance to command a second test flight after the unqualified success of Gemini 3. It would also offer further vindication after unsubstantiated rumours that he had somehow panicked and blown his hatch into the sea after his earlier MR-4 flight aboard *Liberty Bell 7* in July 1961. Despite certain insinuations associated with the loss of that spacecraft, particularly in the 1983 film *The Right Stuff* (based on the novel by Tom Wolfe), Grissom was a consummate and courageous test pilot, and one not given to panic. Had that been the case, he would never have been handed the first Gemini mission by NASA, let alone the crucial maiden flight of an Apollo spacecraft.

In November 1966, Grissom penned a widely syndicated column about his upcoming mission in which he revealed his hopes for the mission and for America's next steps in space. Briefly quoted here, it was published under the title 'Three Times a Command Pilot'.

> In *Liberty Bell 7*, I was a man in a can just along for the ride. *Molly Brown*, bless her heart, was a machine I could maneuver. And now in Apollo 204, Ed White, Roger Chaffee and I will be in a spacecraft designed to go to the moon and back.
>
> Soon I'll be the first United States astronaut to make three flights – one in each of our first three space programs. My upcoming flight is in an Apollo spacecraft which makes my old Mercury *Liberty Bell 7* look something like an early flivver. But in those days we weren't all that concerned about maneuverability. We were out to discover whether man could survive G-forces of lift-off and the environment of space. And we learned that man could survive.[3]

### 'Fire in the Cockpit!'

Not everything was progressing well for the first crewed Apollo mission, and everyone involved was feeling the tension. An unacceptable and frustrating number of problems were causing concern throughout NASA and an increasing level of angst for the crew. Despite these, training for the mission continued, and the launch date of 21 February 1967 still held. All the crew members were professional test pilots accepting of the perils they faced in test-flying a new and untried spacecraft. During a 1966 interview for Associated Press, mission commander Gus

Grissom said he was aware of the dangers. 'If we die,' he stated, 'we want people to accept it. We are in a risky business and we hope that if anything happens to us, it will not delay the program. The conquest of space is worth the risk of life.'[4] His words were to prove sadly prophetic.

An important test of the spacecraft systems, involving the entire countdown sequence, had been scheduled for 27 January. On that day, Apollo spacecraft 012 – the twelfth vehicle in that series designed and constructed by North American Aviation – was mounted in place above the Saturn IB rocket on Launch Pad 34 at Cape Canaveral's Air Force Station, just as it would be the following month, although on this occasion the Saturn was not fuelled. A final, fuelled test would only take place closer to the February launch. On this occasion the crew were taking part in a 'plugs-out' test, or Flight Readiness and Countdown Demonstration Test, which would be carried out with the spacecraft cabin internally pressurized using 100 per cent oxygen. It was not considered a particularly dangerous test of the spacecraft and its systems.

The decision to utilize a pure oxygen cabin environment had been made early in the design of the spacecraft. With weight at a premium, the engineers at North American Aviation had examined every possible way to trim any excess weight. One option they explored was proceeding with a single-gas, pure oxygen environment, which would allow the installation of a far lighter and far less complex Environmental Control System (ECS) than one that produced a mix of 20 per cent oxygen and 80 per cent nitrogen. This would also help mitigate any possible risk of the crew developing the bends as a result of nitrogen being introduced into their blood. Highly pressurized oxygen – as used in this test – always presented something of a combustion risk, but pure oxygen had worked satisfactorily in the Mercury and Gemini spacecraft, and a decision was therefore made to press ahead with this option.

That chilly afternoon on 27 January, the three space-suited crew members climbed aboard the transfer van that would carry them out to the launch pad. By this time, technicians had powered up spacecraft 012, with electric current now surging through several kilometres of coiled, thickly bundled wire that snaked around the floor and wall, and into cavities beneath and above the astronauts' contoured couches. In his later memoir, Gemini astronaut John Young said there were crucial differences between the wiring in the Gemini and Apollo spacecraft. To save on labour costs in maintaining North American's budget, the Apollo spacecraft's wiring had been bundled by machines, and the coating on some of it appeared frayed. Young was concerned about this and numerous other problems, which he put down to North American exploring every possible avenue of using cheaper options. 'I knew it when I saw it,' Young wrote, 'and I saw it in spades in the command module.'[5]

On arrival at Launch Pad 34, the astronauts were transported by elevator up the gantry to the spacecraft level, where they entered what was known as the White Room – a small, enclosed cubicle that provided a protected entry to the spacecraft. They were ready for another long afternoon of checks and tests as they were inserted into the command module and strapped into their respective couches. Once they were settled, the crew plugged into the spacecraft's communications and oxygen systems as they would do in procedural preparation for the actual lift-off.

At 2:42 p.m., pad technicians closed the spacecraft's inner hatch. Unlike the outward-opening hatches used on Mercury and Gemini spacecraft, this brutally heavy hatch would only open inwards, similar to the doors on pressurized commercial aircraft. The hatch was above centrally seated Ed White's head. At one stage of the spacecraft's development, NASA had considered installing explosive bolts in the hatch to allow an emergency egress by the astronauts, but there were fears the hatch might accidentally blow – as happened on Grissom's Mercury mission – so the idea was binned.

Secured by a series of dog-leg latches that had to be operated by ratchets, the hatch would also be held firmly closed by the interior pressure of the cabin, which was higher than the outside atmospheric pressure; again, much like on commercial jet aircraft. Therefore, before the hatch could be opened, venting had to occur until the pressure inside and outside the command module was equal. Then a wrench had to be used to loosen the six bolts securing the hatch. In evacuation simulations, this procedure had never taken less than 90 seconds. As the astronauts had expressed deep dissatisfaction with this cumbersome system, a far simpler, hinged hatch was undergoing development at North American, but it would not be available for this flight, as spacecraft 012 had already been transported to the Cape.

Once technicians had closed and secured the inner hatch, they next secured the outer crew access hatch, and finally locked the booster protection cap into place. In all, three closed hatches were now positioned between the astronauts and the gantry crew. Inside the command module, the crew were busy purging their space suits and the cabin of all gases apart from the 100 per cent oxygen. The cabin's interior was then pressurized to 16.7 psi to simulate normal flight conditions.

Before too long, a series of minor problems began to vex the crew as they worked their way through a lengthy series of checklists. At one stage, Grissom reported a foul odour in the space suit's oxygen supply, which he described as 'a sour smell somewhat like buttermilk'. Then intermittent failures in the communications system between the Operations and Checkout Building and the blockhouse at Pad 34 forced a hold in the simulated count at the 5:40 p.m. mark. Forty minutes later, an exasperated Grissom spoke to ground control, growling, 'How are we going to get to the moon if we can't talk between two or three buildings?' They were probably his last words.[6]

At 06:30:54, a significant surge was recorded in the AC bus 2 voltage readings, indicating a possible short circuit. At the same time, other monitors revealed a sudden spike in the oxygen flow into the crew's space suits, while Ed White's heart and respiration rates suddenly soared. It is believed that a brief electric arc suddenly flared between two bare segments of frayed wire in a small panel situated below the left-hand side of Grissom's couch.

From then on, the cockpit recording is difficult to interpret in places, but eight seconds later one of the astronauts – possibly Chaffee – is shouting what sounds like 'Hey!' Sudden sounds of movement were being monitored within the spacecraft. Grissom undid his harness and knelt on his couch, banging his helmet hard on the upper instrument panel, leaving deep gouges in the top of the helmet. The role he had practised in emergency drills was to lower White's headrest so that White could reach above and behind his left shoulder to actuate the ratchet device that would simultaneously loosen each of the six latches. By this time, flames were rapidly sweeping up the interior wall of the spacecraft, fuelled by the pressurized oxygen environment. At 06:31:06, a voice believed to be White was heard crying out, 'We've got a fire in the cockpit!' as he hastily disconnected his oxygen inlet hose ahead of tackling the inner hatch release.

Roger Chaffee turned up the lights and opened communication links. Ed White could briefly be seen on television monitors inserting the ratchet tool into a slot in the hatch. He suddenly snatched his hands back before reaching out once again as Grissom's hands also came into view, in a desperate attempt to help White with the hatch. Then the TV screens flared and went dead. Meanwhile, the flames, mostly on Grissom's side of the spacecraft, rapidly grew in intensity, melting the oxygen tubes leading into the men's closed helmets, while poisonous gases released by the fire would quickly asphyxiate the three astronauts.

Recent research, combined with the independent findings of some NASA engineers, would indicate that Grissom had also tried to purge the cabin of pressurized oxygen by thrusting his gloved hand through the flames in a vain attempt to activate the cabin dump valves, situated on a shelf over the left-hand equipment bay. There is evidence to suggest that he pressed so hard and violently that the valves bent, although it is doubtful they would have countered the rapidly soaring heat and internal cabin pressure. The temperature inside the spacecraft had escalated to the extent that a number of stainless steel fittings had begun to melt, while molten balls of nylon were dripping onto everything. White's safety harness had already caught fire.

It is known and acknowledged that White put in a superhuman but ultimately futile effort to open the inner hatch, which was not only sealed by internal pressure but expanding in the ferocious heat. There is evidence that he had actually made part of a full turn before being overcome by the deadly fumes.

The last transmission from the spacecraft was Chaffee yelling in despair, 'We have a bad fire!' followed by 'We're burning up!' Then came a sharp, unidentified cry of pain. Seventeen seconds after the first indication of a fire, there was a complete loss of telemetry and communication from the command module.

Meanwhile, the pressure inside the cabin had rapidly escalated to 36 psi, causing a sudden, violent rupture in the spacecraft's hull from an area adjacent to Chaffee's helmet. A firestorm of flames, smoke and debris erupted from the breach into the adjoining White Room, briefly enveloping the exterior of the spacecraft. As recalled by John Tribe, a member of the launch team:

> Within seconds of the first [fire] call, the vehicle basically exploded on the inside. The pressure vessel ruptured and that threw debris through the access doors of the command module that spread across the eighth level [of the gantry tower],

burning technicians, setting fire to the papers on the desk of the pad leader and the flames licked up the side towards the live launch escape rocket that sat on top. The situation was extremely dangerous.

Six [technicians] in pairs of two entered the White Room, which was full of black, toxic smoke. They could not breathe; they could not see. They burned their hands, but they finally got those hatches off. Unfortunately, it was all in vain.[7]

Following the violent hull rupture, the cabin interior was rapidly purged of oxygen and the fire subsided. Just five minutes after the alarm was first raised the booster cover cap had been opened, followed soon after by both the inner and outer hatches. Thick, choking black smoke still clouded the interior of the spacecraft, and nearly five minutes would elapse before it had cleared sufficiently to allow access to the crew members.

It was a nightmare sight. Roger Chaffee was still strapped in, while Ed White's body had collapsed across his seat. Gus Grissom was found lying on his back on the floor of the spacecraft, after his valiant but unsuccessful attempt to purge the cabin of oxygen. In fact the bodies of Grissom and White in their charred space suits were so intertwined that initially it was hard to tell them apart. As soon as humanly possible, physicians Fred Kelly and Alan Harter conducted a brief examination of the occupants, but sadly announced what everyone already knew – that all three astronauts had perished.

It would later be concluded that the three crew members had almost certainly lapsed into unconsciousness some 17 seconds after the first cry of alarm and died from asphyxia caused by inhaling toxic gases from the fire. They had suffered burns of varying seriousness to their bodies, but these would have been survivable had the hatch been opened in time, thanks to the protection afforded by their space suits.

Overnight, the massive, $15 billion American space programme came to a dramatic halt, and would not resume until conclusions had been reached as to the causes of the fire. The most obvious and under-estimated factor was the pure oxygen environment. The day after the fire, a board of inquiry was appointed.

Three months later, the Apollo 204 Review Board, chaired by Langley Research Center director Floyd L. Thompson, issued its preliminary and highly critical report, with complacency given as the root

The fire-gutted interior of the Apollo 1 spacecraft after the crew's bodies had been removed.

cause of the loss of the three astronauts. It was a virtual indictment of almost everyone connected with the flight and the spacecraft, citing as a conclusion, 'Adequate safety precautions were neither established nor observed for this test', and listing the following major direct causes of the accident: pressurized oxygen in a sealed cabin; extensive distribution of combustible materials in the spacecraft; easily damaged electrical wiring; combustible pipes carrying a glycol mixture; poor provision for the crew to escape; and inadequate arrangements for rescuing the crew and getting medical help. The 3,000-page report, which took almost a year to complete, also spoke of 'numerous examples of poor installation, design, and workmanship', as well as 'many deficiencies in design and engineering, manufacture and quality control'.[8] Altogether, it was found that 113 crucial engineering orders had not been carried out when the spacecraft was handed over to NASA. A socket wrench had even been found among the bundled electrical wires in the gutted spacecraft. The cause of the fire, as mentioned earlier, was never established with absolute certainty.

One of the recommendations in the report was the elimination of all easily combustible material from the spacecraft, which would result in the development of new fireproof synthetic fibres. Post-fire, NASA also stipulated that future Apollo space suits had to withstand temperatures in excess of 1,000°F (approx. 540°C). The solution was found in a fabric known as Beta cloth, composed of Teflon-coated glass microfibres, which would be used for the suit's outermost layer.

The report also called for the development and installation of a quick-release emergency hatch; a general tightening of all fire precautions; modifications to the wiring and ECS; the abandonment of a pressurized pure oxygen environment while at ground level; and an emergency venting system that would reduce cabin pressure within seconds. There was also the serious question of complacency endemic throughout NASA, and this had to be addressed with immediate effect.

As one of the three backup crewmen, Walt Cunningham, later reflected: 'Out of the whole mess, North American was to bring forth one of the greatest machines ever built by man. I am convinced that it would not have been possible to reach the moon in only five missions had we not gone through this rebuilding process, which was the inescapable result of the fire on Pad 24.'[9] As recalled by Gerry Griffin, a guidance navigation and control systems officer – later a flight director – for the Apollo missions, 'The flight had been plagued with problems. Apollo 1 was a tragic event and we lost three really good friends, but it may have saved the programme. If we'd had something like that happen on the way to the moon, it probably would have ended.'[10]

### The Wally, Walt and Donn Show

By September 1967, eight months after the pad fire tragedy, a wholly redesigned and redeveloped Apollo command module had been built by North American, incorporating a number of fire precautions and a quick-release hatch which, aided by a compressed-nitrogen cylinder, meant the hatch could be opened within five seconds.

On the question of a pure oxygen environment, which needed to be maintained while the crew was in orbit, a compromise had been reached. On the launch pad, the cabin atmosphere would be maintained at 60 per cent oxygen and 40 per cent nitrogen, and this would change to 100 per cent oxygen during the ascent to orbit. Once in orbit, with the oxygen supply held at just 5 psi, and with almost no convection in microgravity, any fire occurring in space would spread far slower than on the ground, making it something far easier for a trained crew to contain and extinguish. There was also the problem of an astronaut having nitrogen in their body, which might result in dangerous gas bubbles developing during such activities as spacewalks, so breathing pure oxygen in space was a far safer option.

Eventually, NASA began working towards a resumption of the Apollo programme, announcing that when this came about, the backup crew of Wally Schirra, Walt Cunningham and Donn Eisele would fly the first Apollo mission, now designated Apollo 7. Meanwhile, the fatal Apollo 204 mission had also been redesignated as Apollo 1. On 20 September 1968, Schirra stated that this would be his final space mission for NASA, and he was determined to make it a successful flight before he left the space agency. He also declared that on Apollo 7 there would be absolutely no compromises when it came to crew safety.

Between the loss of the Apollo 1 crew and the first manned Apollo mission, several Saturn test flights would be conducted. In a schedule reshuffle, NASA decided to abandon the flights previously designated Apollo missions 2 and 3. Flown but unmanned missions began with Apollo 4 (the first unmanned test of the Saturn V rocket on 9 November 1967), Apollo 5 (a Saturn IB-launched test of a prototype Lunar Excursion Module) and Apollo 6 (a second unmanned Saturn V rocket test). The first post-fire crewed flight would be Apollo 7.

The Apollo 1 backup crew of Donn Eisele (left), Wally Schirra and Walt Cunningham would take over the role of flying the first Apollo mission.

The Saturn IB rocket on the launch pad, ready to send the Apollo 7 spacecraft into Earth orbit.

On the morning of 22 October 1968, 21 months after the launch pad fire, the crew of Apollo 7 were strapped into Command Service Module No. 101, ready to complete the mission originally assigned to their late colleagues. Crucial and often painful lessons had been learned since then, and for his part, Schirra had become an uncompromising and determined commander who would not tolerate any bad practices or sloppy workmanship. He had overseen the development and installation of every system, instrument and fixture on their spacecraft. Schirra saw his and his crew's role as a crucial validation of the Apollo spacecraft that would one day take American astronauts to the Moon.

The mission would lift off from Launch Complex 34, where the Apollo 1 crew had lost their lives. Tensions were obviously high, but there was also confidence in the flight with Wally Schirra at the helm. At one point he expressed concern about strong winds gusting and picking up strength around the launch pad and asked if the flight should be delayed if they gained even more strength, compromising laid-down safety margins. There was every confidence in the 22-storey-high Saturn IB launch vehicle, but there were also restrictions in place, and when the winds hit 35 km/h (22 mph), that limit had been reached. At the same time, the countdown reached the point of ignition and the launch went ahead. Schirra later stated that the launch should have been postponed, as the strong winds could have blown the Saturn rocket back over the beach. 'So someone broke that rule,' he grouched. 'I didn't. I was compromised.'[11]

At 11:03 a.m. EST, the huge Saturn IB rocket lifted off from the launch pad, burning through 2,720 litres (720 gal.) of fuel every second. The rocket carved a straight and true line through the blue Florida skies, with the first stage separating 170 seconds into the flight, following which the second-stage engine took over, reaching full thrust. Eight minutes after lift-off, the crew of Apollo 7 slipped into orbit, leading Schirra to report, 'She's riding like a dream.' Three hours from lift-off, and nearing the end of the second orbit, the astronauts performed the first major manoeuvre of the flight, setting off explosive bolts that separated the command and service module (CSM) from the now fuel-depleted second stage of the rocket.

Unfortunately for Schirra, he developed a head cold on their first day in space, causing him to be a little tetchy when ground control wanted to change the schedule of some tests. Then, to his mounting annoyance, he was asked to conduct a television transmission that day instead of the next. He felt that engineering safety aspects needed to come before disruptive changes to their laid-down flight schedule and wanted to do things in their proper sequence. The important test that day was attempting a rendezvous with their booster. 'I didn't want to mix that up with something else that was not important,' he later complained. Despite his annoyance and obviously suffering from his head cold, Schirra and his crew went through with their live TV show from space, during which he joked around and displayed some of the old 'jolly Wally' everyone had come to expect of the veteran astronaut.

As the flight and tests progressed, it was evident that the redesigned command module they occupied was a superb space vehicle, while the crew performed their duties with great competence and growing confidence in their spacecraft. Schirra, however, was becoming increasingly perturbed at the number of unplanned experiments and tests they were now being asked to perform. On orbit 134 of a planned 163, he finally snapped when yet another unscheduled test was thrust upon them. 'I wish you would find out the idiot's name who thought up this test,' he transmitted back to the mission control team. 'I want to find out, and I want to talk to him personally when I get back down.'[12] After noticing that his crewmates were also becoming frustrated and snappy with the pressure to conduct these new tests, he finally radioed that they would not be doing any further unplanned tests or experiments.

Towards the end of the flight there was further tension between mission control and Schirra when he said that the crew should be able to go through the re-entry phase with their helmets off, arguing that sinus pressure from their colds could possibly result in burst ear drums (although Cunningham and Eisele were not suffering anywhere near as badly as him). It resulted in a terse conversation with the CapCom on duty, fellow Mercury astronaut Deke Slayton.

Schirra won the argument, and was unconcerned about any post-flight ramifications as he had already announced he was leaving NASA after the flight. Having each taken a decongestant tablet an hour beforehand, the crew re-entered without their helmets on and later reported there were no problems with their ears at all. It actually led to crews on later Apollo missions also re-entering without donning their helmets.

The Apollo 7 command module splashed down in the Atlantic Ocean less than 2 km (1¼ mi.) from the planned landing site. It turned upside-down, but this was quickly righted as coloured flotation bags automatically inflated around it. The three exhausted but jubilant astronauts were retrieved by helicopter and deposited on the deck of the main recovery ship, USS *Essex*. The charred command module joined them on deck 43 minutes later.

Flight director Chris Kraft later admitted that while Schirra had been difficult during the mission, he had nothing but praise for the man. 'At times, he gave us a hard time during his flight; technically what he did was superb,' Kraft said. 'On Mercury, Gemini and Apollo, he flew all three and didn't make a mistake. He was a consummate test pilot. The job he did on all three was superb.'[13]

Apollo 7 was Wally Schirra's final space mission. He would retire from NASA as the only astronaut to have flown in all three of America's first space programmes – Mercury, Gemini and Apollo. Neither of his Apollo 7 crewmates would fly into space again: Donn Eisele later served as backup command module pilot for the Apollo 10 mission but resigned from the astronaut office in 1970. He died of a sudden heart attack while on a trip to Japan in 1987. Walt Cunningham subsequently worked in a management role for the follow-on Skylab programme, but after missing out on a prime crew assignment he also decided to resign from NASA, departing in 1971.

The eleven-day mission of Apollo 7 was a perfect test flight of both the Saturn booster and the Apollo spacecraft. The programme's first step towards sending astronauts to the Moon had been accomplished with great competence on the part of both man and machine, and the next mission in the manifest, Apollo 8, received the go-ahead. There was growing confidence that the goal of a manned lunar landing pledged so audaciously by murdered president John F. Kennedy seven years earlier was now tantalizingly within reach and achievable.

### Death of the Soviet Chief Designer

Following the acclaim and political pressure that followed the flights of the Vostok and Voskhod spacecraft, the still unknown Chief Designer, Sergei Korolev, had turned his attention to a new type of space vehicle that might one day carry cosmonauts into orbit, and even to the Moon. In 1962 his design team had also begun work on a massive new booster, called the N-1; a powerful rocket that might carry crews on that lunar voyage.

The new spacecraft, which he called Soyuz (Union), was still under development when, ill and exhausted, he entered a Moscow hospital on 5 January 1966 for what should have been routine intestinal surgery to remove polyps from his rectum following a diagnosis of colon cancer the previous year. Although it was not within his field of expertise, Minister of Health Boris Petrovsky conducted the colon surgery himself at Korolev's insistence. It was not supposed to be a difficult operation, but it went disastrously wrong when Petrovsky came across a fist-sized tumour, and then a large blood vessel suddenly burst open. Korolev already had a badly weakened heart as a legacy of his years in the brutal Gulag, and in the midst of the haemorrhaging it abruptly stopped beating. Frenzied efforts failed to revive him, and Korolev's life came to a sudden end on the operating table. His untimely death on 14 January came just two days after his 59th birthday and would prove a massive, irremediable blow to the Soviet space programme.

It was not until after his death that Korolev's name became widely known to the Soviet people – and the rest of the world – as the man whose rocketry and spacecraft team had placed the first satellite, Sputnik, and the world's first space traveller, Yuri Gagarin, into space. Testimony to his genius lies in the incredible fact that even today, more than five decades after his death, both his rocket and spaceship designs are still flying into space.

While he was alive, Korolev was constantly trying to prevent a number of other competing designers, such as Mikhail Yangel, Vladimir Chelomei and Valentin Glushko, from attempting to impose themselves and their grand schemes into his work. Unfortunately, his death opened the door to a new vanguard of lesser would-be designers, and this was to prove catastrophic for the immediate future of the Soviet space programme. Eventually, 49-year-old Vasily Mishin, Korolev's long-serving colleague and chief deputy, was named as his successor. Almost immediately, Mishin found himself swept up in unrelenting governmental pressure to achieve his predecessor's plans of orbiting the Moon in 1967 and landing a cosmonaut on the lunar surface the following year. He knew all too well that he was inheriting vastly inadequate financing and a fragile and fragmented infrastructure, along with a brutal workload and immediate demands from above to beat a now technologically superior America to the Moon. History records that the first crewed space flight under Mishin's design leadership ended in a catastrophic tragedy.

It is an undeniable fact that the untimely death of Sergei Korolev, and the spectacular failures suffered by his lesser successors, were the undoubted cause of the Soviet Union finally losing the race to the Moon. We will never know what might have happened had Korolev survived his surgery and colon cancer to continue his crucial work as the Soviet space programme's Chief Designer, but his renowned engineering and design brilliance would surely have made the race to the Moon a far more close-run thing.

## A Cosmonaut Is Lost

It had been 25 months since the flight of Pavel Belyayev and Alexei Leonov aboard Voskhod-2 had ended with the safe (if delayed) retrieval of the crew and spacecraft, and during that hiatus the entire programme of America's Gemini missions – ten flights in all – had been successfully completed. By now, American astronauts had convincingly demonstrated that NASA was ready to aim for the Moon. Orbital docking and rendezvous had been achieved and spacewalks performed with varying degrees of success, but with increased capability at each attempt. Astronauts had carried out extended missions to simulate lengthy flights to and from the Moon, and the Soviet Union's lunar plans were being increasingly left in the wake of each successful Gemini mission, completed at a remarkable average of one crewed flight every two months.

Then, on 23 April 1967, came a long-expected announcement from TASS that the Soviet Union was back in space again with the successful launch of Soyuz-1, carrying a single occupant. Colonel Vladimir Komarov had previously commanded the Voskhod-1 mission, and he now became the first Soviet cosmonaut to fly twice into space.

At the time, a lot of interest and speculation centred on the name of the spacecraft, Soyuz-1, as the word *soyuz* translates to 'union'. Was it an indication that a docking was imminent with another Soviet spacecraft? Perhaps there might even be an exchange of crew members while in orbit? As history now reveals, the pundits were correct; even as news of this latest space flight was being announced, the rocket carrying Soyuz-2 was already positioned on the Baikonur launch pad, and a three-man crew of Valery Bykovsky, Yevgeny Khrunov and Alexei Yeliseyev was ready to follow Komarov into orbit and link up with Soyuz-1. The ambitious but hazardous flight plan called for Khrunov and Yeliseyev to exit their spacecraft after docking and spacewalk across to Komarov's Soyuz-1. All three cosmonauts would later land aboard Soyuz-1, while Bykovsky returned as the sole occupant of Soyuz-2. But this ambitious second flight would never take place.

It was later revealed that soon after lifting off, Komarov had begun reporting serious malfunctions with his spacecraft and was working hard on solutions to save the mission. Of principal concern was the failed deployment of one of two solar panels meant to supply power to the spacecraft. Additionally, despite numerous attempts, Komarov had not been able to manually align Soyuz-1 after a failure in the automatic

Soviet cosmonaut
Col. Vladimir
Komarov would
fly the ill-fated
Soyuz-1 mission.

orientation system. The flight was now enveloped in so much serious trouble that the launch of Soyuz-2, planned for the following day, was postponed, with an increasingly strong possibility of being cancelled altogether.

It soon became evident that Komarov would not be able to rectify the spacecraft's technical issues. After 27 hours, all of his valiant efforts to salvage the mission had proved futile, and he was advised to prepare for a premature landing. Komarov tried to align Soyuz-1 for re-entry but was unsuccessful. However, a second attempt, on the nineteenth orbit, worked, and once the braking rockets were fired the spacecraft plunged back into the atmosphere. Minutes later, with the intense heat of re-entry now behind him, Komarov calmly prepared for the small drogue parachute to deploy and drag out the main chute, but then things went horribly wrong. The drogue chute popped out and opened as planned, but the main parachute became stuck in its housing and could not be hauled out. There was, however, a contingency plan in place, and the reserve parachute was deployed. Unfortunately its shroud lines became entangled in those of the drogue chute, and the reserve parachute did not open.

Komarov would have realized he was in serious trouble, plummeting to the ground with nothing to retard his descent capsule's downward plunge. All he could do was sit back and wait for the now inevitable end to his life. Soyuz-1 slammed into the ground at high speed, crushing the descent module into a heap of twisted wreckage just 70 cm (27 in.) high, instantly killing the cosmonaut. The solid-fuel braking rockets, designed to fire moments before a normal touchdown – but completely useless at such tremendous speed – exploded on impact and the pulverized module immediately erupted into flames.

Several years after the incident, graphic film footage of the aftermath was released, showing the still burning, partially buried wreckage of the spacecraft and the desperate attempts of the ground rescue team to extinguish the flames. Once the fire was out and the wreckage had cooled down, Komarov's horribly charred remains were carefully extracted from what remained of Soyuz-1 and flown back to Moscow.

A shocked sadness and bewilderment swept across the Soviet Union and the world as the unexpected news of Komarov's death was released and broadcast. For a nation that had once revelled in the prestige and many triumphant successes of the Soviet space programme, his tragic but unexplained death came as a mighty blow.

Komarov's remains would be inurned in the Kremlin Wall Necropolis on Red Square with full military honours on 26 April 1967, attended by his grieving widow, sombre political figures, a cadre of stunned fellow cosmonauts and thousands of Muscovites in mourning for a lost hero.

One startling fact about the ill-fated flight would remain undisclosed for some years: Komarov's backup for the Soyuz-1 mission had been the world's first spaceman, Yuri Gagarin.[14] After protracted efforts by

A depiction of the Soyuz-1 spacecraft flown by Vladimir Komarov.

him to make a second flight, officials relented enough to allow him this conditional position, although as a national hero they were understandably reluctant to risk his life again. No one could know that less than a year later, on 27 March 1968, Gagarin would be killed in a MiG-15 training flight during his ongoing bid to reclaim his flight status. He too would be inurned in the Kremlin Wall.

Typically, the Soviet space hierarchy had thrown an immediate shroud of secrecy over Komarov's death, which brought forth an ugly, and totally unworthy, crop of rumours about the manner in which he had died. One of these stated that while he was still in orbit, but knowing he would be unable to survive re-entry in his badly crippled spacecraft, his wife Valentina had been transported to the mission control centre in the Crimea to bid an emotional farewell to her doomed husband. Other rumours suggested that Prime Minister Alexei Kosygin had spoken to Komarov, stating the nation's pride in him and his inevitable sacrifice in the name of the Soviet space programme.

Even more alarmingly, stories began to spread of American listening stations in Turkey tuning in as Soyuz-1 plummeted to earth, hearing Komarov's last heart-wrenching, passionate words as he cursed the Soviet government that had ordered him to occupy what was known to be a trouble-plagued spacecraft, launched far too early before the problems had been remedied. Even today these terrible rumours still exist, but any serious observers of human space exploration know them to be utterly incorrect. Even some elementary research into the flight would reveal that the programme's mission control was located in the city of Yevpatoriya in western Crimea, while Valentina Komarov would have been in her Moscow home, more than 1,400 km (870 mi.) to the north – an impossible time factor to overcome. Additionally, contact would have been lost between mission control and the cosmonaut soon after the descent module separated from the other two spacecraft modules. This was a perfectly normal occurrence for a returning spacecraft, in which ionized air engulfing the capsule causes a radio blackout that lasts for several minutes, so there could not possibly have been anyone communicating with Komarov during the re-entry phase.[15]

In 1992 the author corresponded with unflown (now late) cosmonaut Alexander Petrushenko. He had been in the mission control centre and was the last person in contact with Komarov. During their exchange, Petrushenko said a perfectly calm Komarov was reporting that he had completed the final correction manoeuvre prior to re-entry and was

The crew that would have been launched aboard Soyuz-2: Yevgeny Khrunov, commander Valery Bykovsky and Alexei Yeliseyev.

preparing for his premature return. Beyond that, and as anticipated, communication with Komarov had been lost.[16]

Some good news did eventually emerge from this catastrophic event, but once again it was withheld for many decades. Had Soyuz-2 been launched as planned the following day, that crew of Valery Bykovsky, Yevgeny Khrunov and Alexei Yeliseyev would almost certainly have perished as well. Investigations into the horrifying tragedy revealed that Soyuz-2's parachute container carried exactly the same flawed design as that in Soyuz-1, which would likely have resulted in another chute failure and the unavoidable death of anyone on board.

# 6

# EYES ON THE MOON

As history records, one of the most audacious decisions made by NASA in just their seventh year of launching astronauts into space became a reality on 12 November 1968. On that date a decision was made to bring the space agency's lunar landing programme forward by sending the crew of Apollo 8 around the Moon, drawing America tantalizingly close to landing the first – as yet unnamed – human beings on another world.

Had Apollo 8 gone according to schedule, it would not be remembered with quite the same amount of pride and admiration as today. Following the successful Earth-orbital test flight of Apollo 7, the follow-on mission was planned as one that would check out the ungainly-looking lunar module (LM) in Earth orbit. If all went well, another LM would carry two astronauts down to the surface of the Moon after disengaging from the command module in lunar orbit. Although this was a vitally important step in sending astronauts to the Moon, the nation was eagerly anticipating the following Apollo missions that, in just a few months, would carry the first astronauts to the Moon and a landing on its largely unknown surface. It was an exciting prospect, yet another precursory mission spent whirling around Earth, no matter how crucial to NASA's lunar plans, simply did not attract a huge amount of public interest.

The major decision-makers in completely changing the mission goals for Apollo 8 were the then acting administrator of NASA, Thomas Paine (following the retirement of James Webb in October 1968); George Mueller, the associate administrator for manned space flight; and Apollo programme director Samuel Phillips. Several frustrating delays had been severely impacting the mission schedule, and by August 1968, Apollo Spacecraft Program manager George M. Low had found himself with a dilemma. While the command and service module (CSM) had more than adequately proved itself during the Apollo 7 test flight, numerous critical developmental problems were combining to delay the arrival of the first crew-ready LM. A basic prototype had successfully test-flown on the unmanned Apollo 5 mission that January atop a Saturn IB, but the crewed version needed to be as close to operationally perfect as the Grumman engineers, who had designed and developed the landing craft, could make it.

While George Low pondered the fact that the end of the decade was fast approaching, as was Apollo 7 – the first crewed Apollo mission since the tragic launch pad fire that killed the crew of Apollo 1 the previous January – he was also acutely aware of intelligence reports suggesting that the Soviet Union were covertly planning to send a cosmonaut or two around the Moon by the end of 1969 – and could even be working on their own lunar landing craft. Therefore, on 9 August 1968, he met up with Manned Spacecraft Center (MSC) director Robert Gilruth to discuss sending the Apollo 8 CSM around the Moon, to prove that capability, and reassign the LM test flight to Apollo 9. Gilruth was supportive of this initiative and he in turn consulted his top team of Deke Slayton and director of flight operations, Chris Kraft.

Looking back at this situation fifty years on, in 2018 the NASA History Office said that the next step was to seek support for the plan from managers at the Marshall Space Flight Center (MSFC) in Huntsville, Alabama, and the Kennedy Space Center (KSC), as well as NASA headquarters:

> That same afternoon, the four flew to Huntsville and met with MSFC Director Wernher von Braun, KSC Director Kurt Debus, HQ Apollo Program Director Samuel C. Phillips, and several others. By the end of the meeting, the group identified no insurmountable technical obstacles to the lunar mission plan, with the qualification that the Apollo 7 mission in October must be successful. Von Braun was confident the Saturn V would perform safely and Debus believed KSC would be ready to launch in December. They agreed to meet the following week at NASA HQ for a final assessment of the plan and subsequently brief NASA Administrator James E. Webb and Associate

Administrator for Manned Space Flight George E. Mueller. Until those meetings, the group agreed to keep the new plan a secret. The MSC delegation returned to Houston, where Low ended his long day by briefing contractor representatives for the CSM (North American Rockwell) and the LM (Grumman Aerospace) for their input to the plan.

Once this consensus had been reached, the next task for Slayton was to inform the Apollo 8 and 9 crews of this major change to their respective missions.

Slayton had a phone call to make, to Frank Borman, the commander of the planned Apollo 9. Borman and his crewmates were in Downey, California, conducting tests with their CM. Slayton ordered him back to Houston and offered him command of the new circumlunar Apollo 8 mission. Borman immediately accepted. This meant that Borman, along with Command Module Pilot (CMP) James A. Lovell and Lunar Module Pilot (LMP) William A. Anders would fly Apollo 8 on a circumlunar mission without a LM in December 1968. The originally assigned Apollo 8 team of Commander James A. McDivitt, CMP David R. Scott and LMP Russell L. Schweickart would instead fly the Apollo 9 Earth orbital first LM test in early 1969. Swapping the crews made sense, since Schweickart was already a LM expert, while Anders had less LM experience. McDivitt agreed with the swap, deciding to put his crew's experience to good use and be the first to fly the LM.[1]

According to the mission timetable, the crew of Apollo 8 was training to conduct the first test of the spindly lunar module in Earth orbit, but on 19 August NASA announced that it was deleting the LM from the flight due to delays in getting it ready for a December flight. Instead of delaying Apollo 8, Apollo programme director Samuel C. Phillips said NASA was looking at other options for advancing the programme to meet President Kennedy's goal of landing a man on the Moon before the end of the decade. Beginning in early August 1968, senior NASA managers had been contemplating whether it would be possible to send Apollo 8 on a circumlunar mission or even have it enter orbit around the Moon. Once they had explored all the options and consulted everyone associated with the flight, including mission commander Frank Borman, command module pilot Jim Lovell and lunar module pilot Bill Anders, there was a consensus that the new flight agenda should be implemented. The only misnomer in operational crew titles was the one given to Anders, as there would be no lunar module carried on the Apollo 8 mission.

On 12 November 1968, NASA headquarters announced that the Apollo 8 mission would now become an orbital flight around the Moon. This ended weeks of intense internal deliberations within NASA and public speculation about Apollo 8's targeted mission.[2] The launch was provisionally scheduled for 21 December, aiming for ten orbits of the Moon over a total of twenty hours on Christmas Eve and Christmas Day.

Right on the appointed day, 21 December, the launch of Apollo 8 began with almost indescribable violence as a massive explosion of orange flame erupted downwards from the 36-storey Saturn v's five first-stage engines, generating a total of 180 million horsepower. The gigantic rocket howled within its launch pedestal for eight seconds as computer systems ensured all five bell-shaped F-1 engines had ignited and were behaving as they should. Then, with a massive thrust – equivalent to the power of around five hundred fighter jets – the Saturn v laboriously lifted off. Within twelve seconds it had cleared the 122-metre (400 ft) launch tower, after which it tore a raucous path through the Florida skies, trailing a 150-metre (520 ft) trail of flames and fumes.

Following staging, the Apollo 8 spacecraft successfully achieved orbit. Over the next three hours it would complete two circuits of the planet as checks were made on thousands of systems. Word finally arrived that Mission Control was prepared to commit to trans-lunar injection (TLI), a propulsive manoeuvre used to set a spacecraft on a trajectory to the Moon by restarting the attached third-stage rocket. This would accelerate the spacecraft's speed from about 28,000 km/h (17,400 mph) to around 39,200 km/h (24,400 mph), necessary to escape the influence of Earth's gravity. The outward journey to the Moon was relatively uneventful, apart from Frank Borman having a stomach virus that for a time caused him to feel nauseous and feverish.

At 326,200 km (202,716 mi.) from Earth, the crew of Apollo 8 crossed the 'great divide' in interplanetary space, passing beyond Earth's influence into the gravitational pull of the Moon, becoming

The lunar-orbiting crew of Apollo 8. From left: Jim Lovell, Bill Anders and Frank Borman.

the first humans to ever come under the influence of the Moon's gravity. They beamed back fuzzy television pictures of Earth, with commander Borman describing what geographic features could be seen.

On Christmas Eve 1968, Apollo 8 completed a successful burn of its engine and slipped into lunar orbit, with the three crew members gazing in awe at the surface of the Moon 97 km (60 mi.) below as they streaked over the stark, barren plains. 'The view at this altitude, Houston, is tremendous,' Lovell reported. 'There is no trouble picking out features that we learned on the map. The colour looks like a whitish grey, like dirty beach sand with lots of footprints in it. It looks like plaster of Paris. We can see a lot of detail.'[3]

During the fourth orbit of the Moon, the crew's attention was still drawn to the grey, pockmarked lunar surface below as Borman slowly rotated the spacecraft. Suddenly, there was a shout from Bill Anders: 'Oh, my God. Look at that picture over there! There's the Earth coming up. Wow, is that pretty!' He knew he only had a very short time to record the sight of Earth rising over the Moon's horizon and urgently asked for a roll of colour film, which he then loaded into his high-end Swedish-manufactured Hasselblad 500C camera and began snapping away. 'You got it?' Lovell asked. Anders said he had. That image, dubbed *Earthrise*, became one of the most iconic images of the Space Age.[4]

There would be another memorable occasion that will forever be associated with the Christmas Eve lunar mission of Apollo 8. There was a televised programme from the crew, although they were not seen. Instead they pointed the camera down at the surface of the Moon as they began sending Christmas greetings down to Earth. Then, unexpectedly, one by one, the three astronauts began reciting the first ten verses from the Bible's Book of Genesis, starting with Bill Anders reading verses 1 to 4, which begin: 'In the beginning, God created the heavens and the earth'. When he had finished, Jim Lovell took over, reading verses 5 to 8, and finally Frank Borman read verses 9 and 10, concluding

The glorious sight of Earthrise, as photographed by Bill Anders. Although popularly shown with Earth above the Moon, this photograph is in its correct orientation.

the live transmission by saying, 'And from the crew of Apollo 8, we close with good night, good luck, a Merry Christmas, and God bless all of you, all of you on the good Earth.' The Bible reading came as something of a surprise to the normally pragmatic NASA chiefs, but even though they had been caught off-guard, their reaction was the same emotional one that had swept across the nation; it was a moving message of faith, peace, joy, discovery, and the shared spirit of Christmas. Perhaps today's world would frown upon such Gospel readings pertinent to the Christian faith, but for those few precious moments America was united in the sense of awe and profound reflection that the crew of Apollo 8 had relayed to them from lunar orbit. The voices of three American astronauts at Christmas time brought to the world one of the most powerful and meaningful transmissions ever made from the infinity of space.

When it came time for the most critical manoeuvre – the burn that would tear their command module from the grip of lunar gravity – it was orbiting behind the Moon and the crew were out of touch with ground controllers. They knew that if the engine failed to ignite they would be trapped in lunar orbit, and the oxygen supply would eventually run out. Fortunately, the engine fired, and as they regained contact with Mission Control, a jubilant Lovell's first words were, 'Please be informed there is a Santa Claus.'

Towards the end of their journey home, the crew jettisoned the service module portion of their spacecraft and positioned the command module for re-entry, with the blunt heat shield facing the direction of travel. For these and later Apollo crews returning home, the return journey was much faster than the outward one, and they would hit the atmosphere at tremendous speed, close to 40,000 km/h (25,000 mph). To put this in perspective, the distance between a modern jetliner travelling at a regular cruising altitude of around 35,000 ft and the ground would be covered by the astronauts in just one second.

Having survived their fireball re-entry, the crew prepared themselves for the parachute deployment and then the impact with the Pacific waters. All went as planned, and the spacecraft splashed down within 5.5 km (3½ mi.) of USS *Yorktown*. Within minutes, a helicopter was hovering overhead and communicating with the crew aboard Apollo 8. Once the three astronauts were safely aboard *Yorktown*, a team of fifteen doctors spent four hours examining them, eventually reporting they were in excellent health.

There is an unverified story that in the year of Apollo 8, a year that saw the assassination of presidential candidate Robert Kennedy and civil rights activist Martin Luther King Jr, Frank Borman is said to have received a telegram – one of thousands, and sadly discarded – in which a woman simply stated, 'Thank you to the crew of Apollo 8. You saved 1968.'

## Yielding the Moon

Despite strenuous denials, the Soviets had indeed been covertly reaching for the Moon, and the secrecy surrounding these efforts was deeply concerning to NASA, who had no idea if Soviet plans included cosmonauts travelling to the Moon. Back on 8 March 1965, the then administrator of NASA, James Webb, gave testimony on the subject before the Senate Space Committee, telling them:

> We do not know whether they have selected some specific goal, such as a lunar landing, or even a duplication of our Apollo mission . . . There is no evidence that they are building a booster as large as the Saturn V . . . I think the information of most value to the U.S. Government is that they are conducting a very broadly based program, developing every competence necessary to select those missions that they believe will be to their advantage as they develop their competence.[5]

As NASA was later to discover, the Soviet Union did indeed have advanced plans for circumlunar and lunar landing flights very similar to those of Project Apollo. In the latter, the major difference was that only two cosmonauts would be involved instead of NASA's three. While one cosmonaut descended to the surface of the Moon, his companion would remain in lunar orbit. Prior to his death in January 1966, Chief Designer Sergei Korolev had been actively involved in designing a massive new booster called the N-1. This was the rocket he envisioned would launch the LK spacecraft (Lunniy Korabl, translating to 'Lunar Craft') that would land on the Moon.

On separating from the main spacecraft while in lunar orbit, the landing cosmonaut would descend to a height of 110 m (360 ft) above the surface, where he would hover the craft until a safe landing spot could be located. He would then manually fly the LK (originally called

L-1) to that area, ready to abort the landing if the terrain was seen as totally unsuitable and could possibly tip the lunar lander over. Several senior cosmonauts, including veteran spacewalker Alexei Leonov, spent many hours practising the landing in Mi-4 helicopters and inside a simulator of the LK, which was attached to the arm of a centrifuge. He was also teamed with unflown cosmonaut Oleg Makarov, who was in line to pilot the lunar-orbiting main craft.

Much to Leonov's personal frustration, the United States was boldly forging ahead with its lunar landing programme, while Soviet plans were lagging well behind, mostly due to delays with the N-1 launch vehicle. What made it even worse for Leonov was the 14 September 1968 lift-off of the unmanned Zond 5 mission, launched aboard a Proton-K/D rocket and successfully achieving trans-lunar injection. The spacecraft eventually swept around the Moon, carrying tortoises and other biological specimens as well as a full-size mannequin dressed in cosmonaut attire. Leonov knew that Zond 5, a modified version of the Soyuz 7K-L1, had originally been planned to carry two cosmonauts around the Moon, but the failure of two earlier Zond missions had led to fears that the crew might be lost, so caution prevailed. The less cautionary days of the propaganda-seeking Nikita Khrushchev were obviously long gone. The mission was successful, and following its trans-lunar journey, Zond 5 and its occupants splashed down safely in the ocean, retrieved soon after by a Soviet recovery team. Leonov knew that he and Makarov could have been crewing that spacecraft.[6]

Meanwhile, development of the mighty N-1 had continued, creating an extraordinarily powerful rocket 105 m (345 ft) tall. The first test launch took place at the Baikonur Cosmodrome on 21 February 1969 but proved a spectacular failure. Eighty seconds into the ascent, the N-1's thirty first-stage engines shut down prematurely, which brought the rocket crashing back to the ground well away from the launch site. According to Leonov, who was there to witness the launch, it was not a totally unsurprising thing to happen on the rocket's maiden flight, and as he remarked, the rocket had a fatal flaw. 'Its thirty engines were arranged in two concentric circles. When they were all fired up at the same time, a damaging and destabilizing vacuum was created between the two circles. This had not been discovered prior to launch because we had no facilities to test all thirty engines together.'[7]

Leonov then attended the second N-1 launch, but this too turned into a disaster for the Soviet space programme when, just seconds after launch, the rocket was engulfed in a massive fireball and collapsed back onto the launch pad. Two years later, in June 1971, and again in November 1972, two N-1 launches went spectacularly wrong, with both rockets destroyed. Another launch had been tentatively planned, this time carrying an L-3 spacecraft (a Soyuz craft modified for lunar landings), but in May 1974 the N-1/L-3 programme was cancelled.

In a later memoir, Leonov would reveal that nothing could match the deep frustration he felt when news came through of the successful Apollo 8 lunar-orbiting mission, and then when the crew of Apollo 9 performed a masterful test of NASA's lunar module in Earth orbit. 'By this point I knew we were not going to beat them,' Leonov wrote. 'Although no firm dates had been announced, we believed there would be one more Apollo mission before the ultimate goal of lunar landing was attempted. We would never be able to accomplish the lunar missions we had planned on schedule.'[8]

Surprisingly, the fact that the Soviet Union had also constructed their LK-3, an advanced lunar landing craft, would remain a well-hidden secret until 1989, when a group of aerospace engineers from the Massachusetts Institute of Technology (MIT) were being given a tour of the student engineering laboratory at the Moscow Aviation Institute and saw an ungainly looking spacecraft perched in a corner of the building. When they asked their guide what it was, he blithely informed them that this was the actual spacecraft that had been intended to land on the Moon. They quickly took photographs, and once they had returned home, MIT issued a press release about the amazing revelation. On 18 December the *New York Times* ran a front-page article headlined: 'Now Soviets Acknowledge a Moon Race'.[9]

Interestingly, a full-scale engineering model of the LK-3 was shipped from Russia to the Science Museum in London as an exhibition centrepiece in the museum's presentation 'Birth of the Space Age'. It was on open display until the exhibition ended in March 2016.

### *Spider* and *Gumdrop* in Orbit

When the Saturn V rocket assigned to the Apollo 9 mission was transported out to Launch Pad 39A on 3 January 1969, on a seven-hour journey aboard a gigantic crawler tractor, it would be a make-or-break mission in achieving Kennedy's pledge to place astronauts on the Moon by the end of the 1960s. For the first time, all spacecraft components would be launched together – the command and service modules and the fragile-looking Lunar Module (LM), manufactured by contractor Grumman Aircraft. From its inception, the landing vehicle was known as the Lunar Excursion Module, but NASA later decided that the word 'excursion' was a little misleading and it was subsequently dropped from the name. However, the acronym 'LEM' was still useful for transmission purposes and was retained.

As director of flight crew operations, former Mercury astronaut Deke Slayton once explained that, in basic terms, the LM comprised two sections – the descent stage and the ascent stage. 'The descent stage was boxlike and rested on four legs. The ascent stage was the buglike crew cabin. There was a docking port atop the ascent stage and a hatch in the front, through which the crew would exit.'[10] On departure from the Moon, the engine below the ascent module would fire and blast it away from the descent stage, leaving it perched forever on the surface of the Moon.

The Christmas 1968 flight of Apollo 8, sending the first human-tended flight around the Moon, had been a resounding success, and with the LM now ready to fly, Apollo 9 could proceed, much to the relief of George Low and many others. In readiness for the flight, the crew had been permitted to give the CM and LM official code names, by which they would be known during communications with Mission Control – the first time this had been allowed since the Gemini 3 mission. As there were two spacecraft involved, it made sense to allocate distinctive call signs to each of them, so for communication purposes, the LM was dubbed *Spider* and, thanks to its pyramidal shape, the CM became *Gumdrop*.

The launch, originally scheduled for Friday 28 February, was postponed for three days because of colds suffered by the crew, which physicians put down to sheer exhaustion after eighteen-hour training days. On Monday 3 March, all was in readiness, and the mighty Saturn v lifted off at the end of a near-perfect countdown.

Two hours into their ten-day mission, CM *Gumdrop* separated from the s-IVB third-stage rocket, following which it revolved 180 degrees and was propelled back towards the third stage, which housed the now-exposed LM *Spider*. Gently guided by CMP Dave Scott, *Gumdrop* closed in on the s-IVB and nosed into *Spider*, until a firm docking was completed, following which the CSM retreated, now attached to the LM, and the combined spacecraft moved away from the spent third-stage rocket, which would eventually re-enter and burn up over the ocean. The engine at the rear of the CSM was then fired, boosting the joined spacecraft into a higher orbit of around 503 km (313 mi.) altitude, ready to conduct their test schedule over the next few days. They spent the rest of launch day and Tuesday thoroughly checking systems within the command module and conducting tests on the main engine. On the third day of the mission, McDivitt and Schweickart made their way through the connecting tunnel and moved into the lunar module, leaving Scott inside *Gumdrop*. The two astronauts then tested the LM's steering rockets while trying out the 'shirt-sleeve' environment, which allowed them to remove their helmets and conduct a six-minute television show. They fired *Spider*'s main descent engine for six minutes, which created sufficient acceleration to dislodge several chunks of the LM's outer insulation foil skin. The crew noticed this but did not believe it was a matter for concern.

Schweickart had been scheduled to make a spacewalk without the usual life support system from the spacecraft, instead wearing a backpack that would provide him with oxygen, purge his exhaled breath and keep his body cool with a water-circulation switch. Unfortunately, he reported he was suffering from a mild case of nausea and had vomited. NASA would not risk the possibility that he might vomit inside his helmet while on EVA, which could result in his choking to death, and the spacewalk was abandoned. Later, on day four, Schweickart reported he was feeling much better, so part of his EVA was reinstated. During his limited spacewalk, he stepped into a pair of fibreglass 'golden slippers' – foot restraints – mounted on the platform outside the lunar module, where he took numerous photographs of the two spacecraft, Earth and the Moon. His EVA would last just under 38 minutes but was an important test of the cumbersome space suit that would be used later that year by the lunar landing astronauts, exposing it to the vacuum of space for the first time. At the same time, Scott opened the CM hatch and filmed Schweickart on the 'porch' of the lunar module.

On day five, Dave Scott undocked from *Spider* and withdrew 180 km (111 mi.) from the LM. Having been cut loose from the command module, McDivitt and Schweickart jockeyed *Spider* around for four hours, firing both the ascent and descent engines as a major test of their capability, following which they jettisoned the descent module. After 6 hours and 22 minutes of separation, the lunar and command modules rendezvoused and linked up once again, and soon after McDivitt and

Apollo 9 CSM *Gumdrop* and LM *Spider* docked together in Earth orbit. Mission commander Dave Scott was photographed standing in *Gumdrop*'s open hatch by Russell Schweickart.

Schweickart made their way back into the command module. Once all was in readiness, *Spider*'s remaining section – the ascent module – was set adrift, and after sufficient separation had been achieved its engine was remotely fired for more than six minutes until its fuel was fully depleted, sending it soaring into a far higher, elliptical orbit until the influence of Earth's gravity brought it gradually back into the atmosphere, where it was destroyed twelve years later, on 23 October 1981.

The Apollo 9 crew would spend the remaining five days gaining further experience in flying the mothership before preparing for their re-entry. Due to bad weather in the western Atlantic, their landing was delayed by one revolution, until orbit 152, which would deposit them some 772 km (480 mi.) south of the original recovery area, where the weather and seas were far calmer. The prime recovery vessel, aircraft carrier USS *Guadalcanal*, was directed to steam through the night to the new splashdown zone. 'We nailed the touchdown on Apollo 9,' Scott later recalled. 'We came down right on target and splashed down next to the carrier, where there was a chopper ready to winch us from the water and a great meal waiting for us. We stopped over in the Bahamas on the way back to Houston, where our families were waiting for us on the tarmac.'[11]

The Apollo 9 mission was an important step on the way to the Moon, providing a crucial engineering test of the lunar module. Today, the flight unfortunately does not share the spectacular prominence of the later lunar landing missions, but space historians agree that it provided NASA with all the data it needed to proceed to the next phase of Project Apollo – the flight of Apollo 10 to the Moon to simulate a crewed lunar landing.

### Last Step to the Moon

On 14 November 1968, Gemini veteran astronauts Tom Stafford, John Young and Gene Cernan were named to fly to the Moon on the Apollo 10 mission, which would descend to just a few kilometres above the lunar surface in a crucial rehearsal for the mission to follow – Apollo 11, planned as carrying the first crew to actually land and walk on the Moon. A decision to go ahead with that flight would depend heavily on the outcome of Apollo 10.

The flight plan for the eight-day mission of Apollo 10 included virtually everything the next crew would do, except for the actual landing. It called for Stafford and Cernan to enter into their lunar module, leaving Young in the command module, and descend to within 15.6 km (50,000 ft) of the Moon's surface, following which they would fly back to the command module, dock, discard the lunar module and, after two and a half days in lunar orbit, fire their main engine and fly back home. One of the crew's more pleasurable tasks during their training was to give the lunar and command modules communications call signs, and they selected the names *Snoopy* and *Charlie Brown* respectively, thus making the *Peanuts* comic strip characters semi-official mascots for their mission.

The launch had earlier been scheduled for 18 May 1969, and on that day the giant rocket stood poised on Launch Pad 39B – the only time this pad would be used for a Saturn v launch, as Pad 39A was already being prepared for the July launch of Apollo 11. At 12:49 that afternoon, the flight of Apollo 10 began, with the three astronauts setting off on one of the riskiest and most ambitious crewed space missions to that time. Within twelve minutes of lift-off, Mission Control reported that Apollo 10 had achieved a parking orbit around the planet.

Once all of the spacecraft's system had been thoroughly checked, approval was given for the trans-lunar manoeuvre, and the engine on the Saturn's s-ivb third stage roared back into lusty life, propelling Apollo 10 on its path to the Moon. When the rocket's fuel had been exhausted, John Young detached the csm and completed a 180-degree turn so it was facing the opening at the top of the rocket, where lm *Snoopy* was waiting to be freed. Young carefully docked with the lm and dragged it out of its nesting place, after which he swung back around and headed their spacecraft on a steady path to the Moon.

The outward journey was relatively uneventful, with the crew taking part in transmitting live television images of themselves and the receding Earth to those back home. After three days, they successfully fired their main engine to send them into their first, elliptical orbit of the Moon. 'This engine is just beautiful,' remarked Cernan. The engine would be fired once again to attain the desired circular orbit of 111 km (69 mi.).

On their ninth lunar orbit, wearing soft overalls, Stafford and Cernan floated through the tunnel connecting the lunar and command

A step closer to a lunar landing would be made by the crew of Apollo 10. From left: Gene Cernan, Tom Stafford and John Young.

modules and began the long process of powering up the lunar lander. Three orbits later, everything was ready and they were given the 'go' for the undocking. John Young was in charge of the separation procedure. The docking latches snapped open and Young carefully backed *Charlie Brown* away from *Snoopy*. Once the LM astronauts had completed their exhaustive checklist, it was time to fire the engine on *Snoopy*'s descent stage, and Young bid his colleagues a temporary farewell. 'Adios. We'll see you back in about six hours.' Cernan responded, 'Have a good time while we're gone.'

The descent engine fired perfectly for the planned 59 seconds and began taking them down to a much lower lunar orbit, as Cernan later recalled in his memoir, *The Last Man on the Moon*:

> Down we swooped, lower and lower, and the Moon ceased being a big gray ball, flattened out, and gave us a horizon. It now almost looked as if we were flying over an Arizona desert, but no desert has ever had such terrain. Those crater walls we had once looked down into now took on the distinct and menacing look of an onrushing mountain range, for it seemed like we were below their peaks. Fields of boulders grew in size and shadows lengthened below canyon walls that grew on both sides, their flat tops appearing to be high above our little flying bug.[12]

As they descended, Stafford and Cernan were also astonished to witness the glory of an Earthrise, a sight that had earlier awed the crew of Apollo 8. Once they had settled into their low orbit they began their assigned tasks, one of the most important of which was to photograph the likely landing site for Apollo 11 in the Sea of Tranquillity. They also tested the landing radar while keeping a constant eye on *Snoopy*'s computer, ready to seize on any problems that might affect the following Apollo mission as it duplicated their efforts prior to the actual landing.

All too soon it was time to begin preparing to return to a rendezvous with Young aboard *Charlie Brown*. The descent stage of the LM was jettisoned, and they made ready to fire the ascent engine when suddenly everything went haywire. Stafford and Cernan found themselves cartwheeling, bouncing and diving as *Snoopy* went completely crazy. Their radar, meant to guide them towards the command module, instead locked on to a bigger target – the Moon. They were in serious trouble, and over the next alarming fifteen seconds a few choice words made their way through their open microphone as they worked to solve the problem. Despite being tossed around, Stafford managed to shut down the computers and took over manual control. It was later revealed that had the loss of control continued for just another two seconds, they would likely have crashed into the Moon.

Stafford told an alarmed Mission Control that the situation had been resolved, and despite what had just occurred they were still on schedule and had fired the engine on their ascent module. Five and a half hours after undocking from *Charlie Brown*, *Snoopy* was once again closing in, and the link-up went well, with the docking latches snapping together. As they passed behind the Moon again, *Snoopy* was jettisoned, and the three astronauts, two of whom were totally exhausted, settled back and slept for nine hours.

After 31 lunar orbits, the CSM's main engine fired perfectly, and the crew of Apollo 10 were on their way home – mission all but accomplished. Nearing re-entry 55 hours later, the service module was also jettisoned, and the command module rotated so that the blunt end of the spacecraft was facing into the atmosphere. All went well with their re-entry, and *Charlie Brown* splashed down under its main chutes, close to USS *Princeton*, after an epic journey lasting 8 days, 3 minutes and 23 seconds.

Across the United States there was jubilation, and nowhere more than throughout NASA. Despite the computer problems on leaving the Moon – quickly identified and rectified – the lunar landing programme could continue to a landing on the next flight. The time had now come to make way for the big one: Apollo 11.

# 7

# ONE GIANT LEAP

O n the evening of 9 January 1969, eight years to the month since Alan Shepard had been named to command America's first human space flight, NASA formally announced the crew for the Apollo 11 mission, planned for July of that year. It would be the fifth crewed Apollo mission, and if no serious problems were encountered on the two planned lead-up missions – Apollo 9 and 10 – then the crew of Apollo 11 would attempt the first landing on the Moon. It meant that the late president John F. Kennedy's pledge of achieving this monumental goal before the end of the decade was still well within reach. The following morning, NASA officials introduced the Apollo 11 crew during a press conference at the Manned Spacecraft Center in Houston.

According to the NASA Manned Spacecraft Center's *Roundup* magazine of 24 January 1969, even though Apollo 11 was considered as the earliest possible mission in the Apollo programme to attempt a landing on the Moon, it was deemed possible that either of the two preceding missions, Apollo 9 and 10, could demonstrate a need to fly 'an alternate mission' on Apollo 11, 'thus moving the lunar landing to a later flight'.[1]

The selection of any Apollo crew, and likewise any Gemini crew, was not an arbitrary thing, and many factors were taken into consideration. The process had its roots in October 1963 when grounded Mercury astronaut Deke Slayton was named assistant director of flight crew operations, in addition to his job managing the astronaut office. This also tasked him with sole responsibility for selecting crews to fly

*Opposite:* 16 July 1969: the Saturn V lifts off carrying the crew of Apollo 11 on their journey to the Moon.

The three NASA astronauts selected to crew the Apollo 11 lunar landing mission. From left: Neil Armstrong, Michael Collins and Edwin ('Buzz') Aldrin.

missions, although his choices would have to be approved by NASA headquarters. Beginning at Gemini 3, and with original commander Alan Shepard unavailable due to a chronic illness, Slayton was going to select Gus Grissom as mission commander along with Frank Borman. However, he soon realized that the two had strong personalities that were incompatible, which could prove disastrous if they were thrown together for any length of time within a spacecraft limiting them to the size of the front seat of a Volkswagen. He prudently moved Borman to a later mission and replaced him with John Young, who subsequently formed a superb, harmonious crew with Grissom.

Slayton now selected his crews based primarily on their work ethic, skills and talents appropriate to the aims of each mission, and that key quality of compatibility. He then devised a long-standing rule that a crew must first perform a backup role for a mission. Assuming this worked well, they would miss the next two flights and then assume the role of prime crew on the third. It worked well in Project Gemini, and for Apollo as well, although some individual changes were made through illness, injury or retirement.

The rotation system meant that Neil Armstrong, Buzz Aldrin and Mike Collins were named as the prime crew for Apollo 11, although the original lunar module pilot (LMP), Fred Haise, had been reassigned to another crew and Collins was brought in to replace him, but as command module pilot (CMP). Aldrin then changed from CMP to LMP and now had a shot at walking on the Moon. Although their flight was pencilled in as the potential first landing crew, it was by no means a sure thing that Apollo 11 would fulfil that role, as much could go wrong before then.

## Crewing Apollo 11

Each of the three crew members named had flown a single mission in the Gemini series. Neil Alden Armstrong was announced as mission commander for Apollo 11. A former X-15 civilian research pilot, he was born on 5 August 1930 in Wapakoneta, Ohio, and was married to Janet Shearon, formerly of Illinois. They had two sons, Eric and Mark, but sadly their daughter Karen died in infancy. As a Navy pilot he had flown 78 combat missions during the Korean War, being awarded three Air Medals. Post-war, he resumed his studies at Purdue University, Indiana, and received his master of science degree in aeronautical engineering in January 1955. Following his graduation from Purdue, Armstrong took up a position with the Lewis Flight Propulsion Laboratory, part of the National Advisory Committee for Aeronautics (NACA). The laboratory was renamed the NASA Lewis Research Center in 1958 and was further renamed the NASA John H. Glenn Research Center in 1999, after the pioneering Mercury astronaut. Here he would test-fly a vast number of high-performance aircraft, including the Bell X-1B, and in late 1958 was one of the first three pilots selected for the NACA X-15 rocket plane programme. From 1960 to 1962 he made seven flights in the winged spacecraft, although he never managed to cross the boundary of space, recognized by the U.S. Air Force as 50 mi., or 80 km, altitude.[2]

For a time, Armstrong was involved in the U.S. Air Force's X-20 Dyna-Soar (Dynamic Soaring) programme. This was a project designed to develop a delta-winged spaceplane that could be used for a number of military applications and reconnaissance missions. The triangular delta wing – named after the Greek uppercase symbol for the letter delta (Δ) would later be adapted for use on the space shuttle.

The matt black Dyna-Soar would have been launched atop a Titan IIIC rocket and achieve Earth orbit, and once its mission had been accomplished the pilot would return to a glided touchdown on a runway, much in the same way as the later space shuttle. Due to the rapidly mounting high cost of the programme and its questionable military potential, the x-20 Dyna-Soar programme was cancelled in December 1963.

In 1962, prior to the abandonment of the x-20 project, Neil Armstrong applied to NASA for selection and training as an astronaut, and was accepted as part of the nine-strong Group 2 contingent. He first served as backup mission commander for Gemini 5 along with pilot Elliot See, and received his first flight assignment as command pilot of Gemini 8, with pilot David Scott. They were launched into orbit on 16 March 1966 and would accomplish the first docking with another spacecraft after rendezvousing with the unmanned, orbiting ATV. Then a faulty thruster in their spacecraft caused the assembly to begin spinning wildly, compelling the crew to undock from the Agena. Once they had brought the Gemini craft under control, they were ordered to abort the remainder of the mission and re-enter, which they accomplished safely.

Armstrong's next assignment was as backup commander for Gemini 11, and once that mission had flown he was named commander of the backup crew for Apollo 8. The crew rotation system then in place meant that with the prime crew successfully flying the Apollo 8 mission, Armstrong was confirmed as the commander of the prime crew for Apollo 11.

Named as Armstrong's lunar module pilot on Apollo 11 was Edwin Eugene 'Buzz' Aldrin Jr. Born into a military family in New Jersey on 20 January 1930, he was the son of an aviation manager and former Army Air Corps pilot, while his mother, Marion, was (remarkably) born with the surname Moon. Continuing a long military tradition in his family, Aldrin was accepted into West Point, the U.S. Military Academy, graduating with a bachelor's degree in 1951. He was then commissioned into the U.S. Air Force and posted to Korea, where he flew 66 combat missions in two years. After serving three years as a general's aide and flight instructor, Aldrin, now married to Joan Archer, enrolled at MIT in 1959, where he was awarded his doctorate in aeronautics and astronautics in 1963, for which he wrote a thesis on rendezvous techniques involving manned orbiting vehicles. The previous year, he had applied to NASA, hoping to be included in their second astronaut intake, but missed out as he was not a qualified test pilot.[3] By this time he and Joan had three children: Janice, James and Andrew. When a third astronaut intake was announced, Aldrin applied, and in October 1963 was accepted as one of fourteen new NASA astronauts.

After three years' training, Aldrin received his first mission assignment as backup pilot for Gemini 10, together with flight commander Jim Lovell, who had earlier flown on Gemini 7. Under the existing crew rotation system then in place, this meant the backup crew would miss two flights and then become prime crew on the subsequent mission – theoretically Gemini 13 – but that flight did not exist. It was a huge disappointment for Aldrin, as the Gemini series of flights would end with Gemini 12, meaning he would miss out on flying any Gemini mission. Then fate stepped in, when the prime crew of Gemini 9, Elliot See and Charles Bassett, died in an aircraft accident in St Louis, Missouri, in February 1966. This meant the backup crew of Tom Stafford and Gene Cernan became the prime crew, and in the subsequent crew reshuffle, Lovell and Aldrin became their backups, which moved them into the position of prime crew on the final two-man mission, Gemini 12.[4]

The two astronauts completed a highly successful mission, the final flight in the Gemini series. During their four-day mission, launched on 11 November 1966, Aldrin completed NASA's first truly successful spacewalks, spending 5 hours and 30 minutes on EVA activities. His background in orbital rendezvous and his masterful EVA experience on Gemini 12 combined to place him on the Apollo 11 mission as lunar module pilot alongside the former X-15 pilot and Gemini 8 commander Neil Armstrong.

Michael Collins was born on 31 October 1930 in Rome, Italy, the younger son of U.S. citizens Major General James Collins, a military attaché, and Virginia Collins. In fact, all three Apollo 11 crew members were born in 1930. After the family moved back to the United States, Collins entered the military academy at West Point, graduating in 1952. He then opted to join the Air Force, and in 1957 married Patricia Finnegan from Boston, Massachusetts. They would have three children: Kathy, Ann and Michael Jr. Following a number of postings and promotions he was assigned in August 1960 to the Air Force's Experimental Flight Test Pilot School in Class 60C, where he attended classroom briefings, flew numerous aircraft types and did performance evaluations. One of his instructors was future astronaut Tom Stafford, as were two of his classmates, Frank Borman and Jim Irwin. Having completed the course, Collins was transferred to Edwards Air Force Base, California, to test new fighter aircraft. In 1962 he was accepted into the Aerospace Research Pilot School, also located at Edwards, training for the possibility of future military space flights. He completed this training in

May 1963. He then stayed on at Edwards as a flight test officer until his application to join NASA's astronaut ranks was accepted, joining the space agency's third group of astronauts, one of whom was his future Apollo 11 crewmate Buzz Aldrin.

After serving as backup pilot for the Gemini 7 mission along with John Young, who had flown on the first Gemini mission with Gus Grissom, the two men were assigned to the Gemini 10 flight. Their mission began with a Kennedy Space Center launch on 18 July 1966, and on this mission they rendezvoused and docked with another ATV. After undocking, they chased after a second ATV and once again performed a flawless rendezvous, following which Collins made two spacewalks.

In September 1966, Collins was assigned to the backup crew for the second crewed Apollo mission as lunar module pilot, along with mission commander Frank Borman and command module pilot Tom Stafford, but three months later he was reassigned to the third crewed Apollo mission – the first to be launched atop the massive Saturn v rocket. The loss of the Apollo 1 crew in January 1967 resulted in several crew changes, and in November that year he was announced as command module pilot on the Apollo 8 crew, along with mission commander Frank Borman and LMP Bill Anders. In 1968, following trouble with his legs, Collins consulted a physician and was diagnosed with a cervical disc herniation, requiring surgery to fuse two vertebrae. As a consequence he was removed from the Apollo 8 crew and replaced by Jim Lovell. Following his complete recovery, he was available for selection to another crew.[5] Deke Slayton regretted that Collins had missed out on the lunar-orbiting Apollo 8 mission and felt that he deserved a seat on the first available mission. The CMP seat he got was on Apollo 11.

### First on the Moon

It was Wednesday, 16 July 1969, and over a million excited people had converged on Cape Kennedy, camping out in cars, caravans and tents to witness the launch of the mighty Saturn v rocket setting Apollo 11 on a course to the Moon. Among those gathered in the VIP stand was America's 'Lone Eagle', Charles Lindbergh, the first person to fly solo across the Atlantic. Thousands of journalists and camera and film people from all over the globe were busy getting ready and setting up their equipment, and as the countdown wound down that day, an estimated 1 billion people were watching on their television sets all around the world. As journalist John Mansfield later wrote of that historic event:

> Exactly on schedule, at 8.9 seconds before 9.32 in Florida, the five engines of the first stage of the Saturn v were ignited: with a shattering roar the great moon rocket built up its 7,600,000 lb. thrust and at 9.32 precisely a relieved but proud voice from Mission Control told the world 'We have lift-off!' At that moment the service arms of the mobile launch tower retracted, and a deluge of 50,000 gallons of water a minute began to cool the tower to stop it melting under the scorching blast of the rocket flame. Below the rocket a specially built flame deflector in the flame trench received the full impact of the heat of lift-off. Its refractory concrete and volcanic ash surface turned to glass, despite a colossal water dousing.[6]

Once it had been drained of all its fuel, the first stage of the Saturn v separated minutes after launch, at which point the second stage ignited, propelling Apollo 11 closer to Earth orbit. When depleted of fuel, that stage was also subsequently discarded and the third stage kicked in. Soon after, the crew reported a successful insertion into orbit.

Having completed one and a half circuits of the planet, during which time the crew and ground control ran meticulous verification checks on the spacecraft and its systems, the third stage of the Saturn v was fired once again, boosting Armstrong, Aldrin and Collins out of Earth orbit and onto a trajectory to the Moon, now travelling at 44,000 km/h (27,300 mph). Soon after, command module *Columbia* and its service module (forming the composite CSM) were separated from the third stage and, with Mike Collins at the controls, performed a 180-degree turn. The CSM then moved back towards the third stage, where lunar module *Eagle* was stowed in a folded configuration. Collins slowly edged into the docking collar of *Eagle*, and once a firm contact had been confirmed he gently withdrew the LM from the spent rocket casing. With this successfully accomplished, Apollo 11 headed for the Moon. The flight continued to proceed satisfactorily and on schedule, and the crew settled back for a well-deserved sleep. The following day, a minor course correction was carried out and the three astronauts took part in a televised broadcast as Earth slowly diminished behind them.

After travelling through space for three days, the spacecraft's main engine was fired for close to six minutes, slowing the backwards-facing spacecraft sufficiently to be captured by lunar gravity. A little over four hours later, another short burst of the engine stabilized the craft, placing it in a near-circular orbit of the Moon.

Once everything was ready, Armstrong and Aldrin said their farewells to Michael Collins, who would remain on board CM *Columbia* as the sole astronaut, and made their way into the cramped confines of *Eagle*, closing the hatch behind them. Once everything had been checked, they powered up the LM's systems and extended the folded landing legs, following which they undocked from *Columbia*, and Collins withdrew to a safe distance. Now flying feet first and head down, Armstrong and Aldrin fired the LM's descent engine, and they were finally on their way to the lunar surface.

It had all gone so smoothly and so far without a hitch, but as the two watchful astronauts descended upside down to the Moon's surface, they were suddenly distracted when alarms began repeating the numbers '12 02, 12 02'. There were several 12 02 alarms and one 12 01 alarm sounding during the powered descent: essentially the computer letting the crew know that it was overly busy and wanted to restart the programme. While the computer always rectified itself, there was still cause for concern, with the word 'abort' on many minds. Fortunately, many of the flight controllers in Mission Control had already encountered this same anomaly during simulations the previous month, and guidance officer Steve Bales (who could have called for an abort of the landing) knew exactly what the alarm meant and allowed the landing attempt to proceed. Each time the alarm sounded, Bales calmly informed Armstrong and Aldrin they were still 'go' for the landing.

The crew had also noted that, due to a slight navigational error and a faster-than-intended descent speed, they would overshoot the intended landing site by around 6.5 km (4 mi.), and were headed instead for a massive crater filled with boulders. They knew that if *Eagle* landed on a pronounced tilt, they might be unable to launch back into lunar orbit. They had to find somewhere more suitable, and meanwhile the lunar module was slowly being drained of precious fuel. As Armstrong later recalled:

When I was finally able to look out we were so low that we couldn't see far enough to identify any significant landmarks.

There was a large, impressive crater which turned out to be West Crater, but we couldn't be sure of that at the time. As we neared the surface we considered landing short of that crater, and that seemed to be where the automatic system was taking us. But as we dropped below a thousand feet, it was quite obvious that the system was attempting to land in an undesirable area in a boulder field surrounding the crater. I was surprised by the size of these boulders; some of them were as big as small motorcars. And it seemed at the time that we were coming up on them pretty fast; of course the clock runs at about triple speed in such a situation.[7]

Taking over manual control, Armstrong's experience and training now came into play as he coolly manoeuvred away from the crater with a subtle touch honed by years of flying some of the fastest jets in the sky. As they reported every move back to Houston, the men's voices remained confident, but the problem remained. No place to land. Huge rocks and surface debris strewn everywhere. And precious fuel was being depleted, also becoming a critical factor.

Aldrin continued calling out navigation data to Armstrong in a steady and clear voice. 'Thirty feet, faint shadow.' Mission Control in Houston issued a warning: 'Thirty seconds [of fuel remaining].'

Then came the call that an anxious Mission Control had been hoping to hear from Aldrin as long probes dangling beneath the LM touched the lunar surface: 'Contact light! Okay, engine stop. Descent engine command override off.' Armstrong gently touched down in the Sea of Tranquillity. He then took a deep breath and reported in. 'Houston,' he declared, 'Tranquility Base here. The *Eagle* has landed.'

It had been a close thing, with only a few seconds of fuel remaining before the thrusters cut out. Had that happened, *Eagle* would have plummeted to the surface. 'Roger, Twan . . . Tranquility, we copy you on the ground,' was the relieved call from a tongue-tied fellow astronaut Charlie Duke in Mission Control. 'You got a bunch of guys about to turn blue. We're breathing again. Thanks a lot.' The two astronauts exchanged congratulations before quickly preparing the LM to launch off the surface in the event of an emergency, but everything seemed fine.

Neil Armstrong's landing on the Moon was actually performed just a little too gently. The lunar module was specifically built with light-weight aluminium honeycomb struts designed to collapse and crush on landing, thus absorbing the shock. Procedurally, Armstrong was supposed to cut the engines when *Eagle* was a few feet above the lunar surface, but instead he set down too gently, and the legs never compressed. This left the cabin of the lunar module several feet higher than intended.

The flight schedule now called for the astronauts to have a four-hour rest period, but Armstrong would have none of that. Everything had gone according to schedule, the LM was in good shape, and both astronauts were far from sleepy. He requested an earlier exit and lunar excursion, which Mission Control supported.

There were still numerous checks to be carried out and a host of suit-up preparations to perform before the EVA could begin. It was a well-rehearsed procedure, and nothing could be overlooked in their eagerness to step onto the Moon. Eventually, 6 hours and 21 minutes after *Eagle* had settled on the surface of the Moon, and having depressurized the lunar module, Armstrong opened the inward-swinging spacecraft hatch. Right around the world, families sat glued to their television sets, breathless with anticipation, pride and not a little anxiety as Armstrong moved down the slender LM ladder, activating a TV camera along the way. The first ghostly, grainy black-and-white images – for a short time inverted – suddenly emerged on screens around the world, showing a space-suited man cautiously descending from that ungainly looking machine known as *Eagle*.

At 10:56 p.m., U.S. EDT, Neil Armstrong reported that he was on the footpad. The bottom step of the ladder was much higher than planned, which meant that the astronauts had to perform a little slow-motion jump down to the footpad – not all that difficult in one-sixth gravity. Armstrong noted as he looked around that the footpads were only depressed an inch or two into the lunar dust, which he described as 'very fine-grained and powdery'.

Armstrong then said he was going to step off the footpad, and the world held its collective breath. Gripping the ladder in his right glove, Armstrong extended his left foot and firmly planted the first human footprint into the surface of the Moon. Then he uttered those immortal words: 'That's one small step for [a] man; one giant leap for mankind.'

There was jubilation around the world as people sat or stood mesmerized by what they were witnessing, watching Armstrong venture further afield, collecting some contingency samples of lunar soil, which he stowed in a pocket in the left leg of his space suit. He was joined on the surface some twenty minutes later by fellow moonwalker Buzz Aldrin, who marvelled at what he saw around him, calling it 'magnificent desolation'.

There were some ceremonial duties they had to perform, the first of which was unveiling a plaque mounted on one of *Eagle*'s legs, which read, 'Here men from the planet Earth first set foot upon the Moon, July 1969, AD. We came in peace for all mankind.' They also planted an American flag in the lunar regolith. Both men reported little difficulty in moving about in the one-sixth gravity, walking warily at first, but as their confidence grew they jumped, bounced and kangaroo-hopped in front of *Eagle*. Meanwhile, they were scooping up dust and rock samples and setting up experiments that would relay data back to Earth.

At one stage, the two moonwalking astronauts were asked to stand by for a telephone message from President Richard Nixon, in which he told Armstrong and Aldrin that through what they had done, 'the heavens have become a part of man's world. And as you talk to us from

Leaving a footprint on the Moon.

the Sea of Tranquility, it inspires us to redouble our efforts to bring peace and tranquility to Earth. For one priceless moment in the whole history of man, all the people on this Earth are truly one: one in their pride in what you have done, and one in our prayers that you will return safely to Earth.'

Once Aldrin had finished setting up his final experiments and Armstrong had gathered up numerous samples, they deployed a passive seismometer to record possible moonquakes and a laser ranging reflector to record the exact distance between the surface of the Moon and that of Earth, before it was time to return to *Eagle*. After hoisting their samples into the LM, Aldrin climbed back in, followed by Armstrong, who closed and sealed the hatch behind him. Altogether they had spent just over two and a half hours conducting their lunar EVA, travelling a total distance of around a kilometre (3,300 ft) as they moved around. Armstrong once wandered as far as 60 m (200 ft) from *Eagle* to visit a large crater. They would be bringing back 21.55 kg (47½ lb) of lunar rocks and soil.

Once safely inside *Eagle* the weary but elated astronauts repressurized the craft, stowed their gear and settled down to a fitful sleep – Armstrong in a crude hammock he had strung across the interior, Aldrin curled up on the floor. Upon waking five hours later, they began making preparations for the return to Mike Collins aboard *Columbia*.

Right on schedule, the ascent engine of *Eagle* fired, lifting it clear of the descent stage platform, which was left behind. It was a speedy departure; within seconds they were high above the lunar surface and ready to execute the rendezvous process with *Columbia*. In the final stages of the rendezvous, Collins was peering out of the right-hand window and could see *Eagle* approaching quite plainly. His task now was to guide *Columbia*'s docking probe into a socket at the bottom of a drogue on the LM, at which time three latches in the probe's head would lock the two modules together. At 5:40 p.m. EDT, Mission Control began referring to the joined spacecraft as Apollo 11, indicating a successful docking had occurred between the two vehicles. 'It's nice to have company,' Collins glibly remarked.[8]

Once the two moonwalkers had removed and cleaned their space suits and other gear of most of the clinging Moon dust, they began

*Opposite:* One of the most seen and iconic images of the 20th century is this photograph of Buzz Aldrin on the Moon, with Neil Armstrong reflected in his visor.

LM *Eagle* on its return journey to a rendezvous with CM *Columbia*.

transferring these and their lunar samples and other equipment into the CM. Four hours after the docking, both hatches between the two craft had been sealed, and a series of small explosive charges surrounding the docking ring were fired. *Eagle* slowly drifted away, to remain in lunar orbit for a few weeks before finally crashing onto the Moon at a location that still remains elusive, despite the best efforts of many researchers.

Later, as they swung around the far side of the Moon, Collins fired the service propulsion system (SPS) single engine that would set them on a course back to Earth, two and a half days away. All that was needed on the return journey was a planned mid-course correction and to jettison the service module.

Eight days, 3 hours, 18 minutes and 35 seconds after it had begun with a launch from Cape Kennedy, the Apollo 11 command module *Columbia* splashed down in the Pacific Ocean about 1,480 km (920 mi.) southwest of Hawaii and just 19 km (12 mi.) from the prime recovery ship, USS *Hornet*. Moments after the splashdown, some Navy frogmen had dropped from a waiting helicopter to secure a flotation ring around the bobbing spacecraft to keep it upright and prevent it from possibly sinking.

Once the okay was given, the crew opened *Columbia*'s hatch from the inside and were handed three biological isolation garments, for the unlikely possibility that they had brought back any toxic lunar bacteria. When they emerged, the three men were also scrubbed down with a povidine-iodine solution. They were then transported by helicopter to the *Hornet*, where they would remain inside a mobile medical quarantine facility (basically a modified Airstream caravan). After being greeted through a window by President Nixon, the quarantine van containing the astronauts was whisked back to the United States by jet transport, where the men would remain for three weeks before being cleared by the doctors and released to a joyous reunion with their families.

President Kennedy's 1961 pledge to his nation had finally been fulfilled, and mankind's greatest-ever scientific undertaking had been accomplished. As President Richard Nixon said in personally greeting the Apollo 11 astronauts following their return to Earth: 'This is the greatest week in the history of the world since the creation.' While that statement is arguably an example of gross hyperbole, what cannot be denied is that in July 1969, for the first time in the entire history of humankind, two people from the planet Earth had walked on the surface of another world.

## Struck by Lightning

NASA's second crewed lunar landing mission came perilously close to being aborted shortly after lift-off. With an all-Navy crew aboard, Apollo 12 was launched on 14 November 1969 from Launch Complex 39A at Kennedy Space Center, Florida, and began its ascent in the midst of rain flurries and dark storm clouds. The weather had no real impact on the determination to proceed with the launch, although the presence of President Nixon in the VIP stand may have contributed to the decision to press ahead.

Then, at 36.5 seconds, and again at 52 seconds, a major electrical disturbance was caused by two totally unexpected lightning strikes. In the first incident there was a cloud-to-ground lightning discharge, which set off warning lights and alarms in the spacecraft's crew compartment. Communications were interrupted, instruments and clocks

The crew of Apollo 12: Pete Conrad, Dick Gordon and Alan Bean.

went haywire, and all three fuel cells disconnected. The second lightning strike stayed in the cloud and had no ground contact, but it caused the Saturn v rocket's navigation system to shut down, and all the circuit breakers in the spacecraft were flashing red. 'I don't know what happened here,' mission commander Pete Conrad reported to Mission Control. 'We had everything in the world drop out.'

According to a post-flight incident report, NASA's Marshall Space Flight Center stated: 'They launched the vehicle into a weakly electrified environment associated with a weak cold front passing right over the space center. Analysis later showed that the cold front's current, though too weak to produce natural lightning, was strong enough for the rocket and its ionized, electrically conductive exhaust plume to produce a charge and generate two lightning strikes.'[9] On the ground at Mission Control in Houston, a set of nonsensical numbers had appeared on flight controller John Aaron's console monitor. He quickly realized that the disrupted readout on his screen resembled one he had seen during an uncrewed flight simulation, the data similarly scrambled by a voltage interruption. He quickly recommended to flight director Gerry Griffin that the crew switch the signal conditioning equipment to auxiliary, which he said would reset the system. Fortunately, LMP Alan Bean knew exactly what that meant and where that switch was located and was able to realign the inertial platform manually. Soon after the system was reset, the fuel cells came back online, and John Aaron once again had good data back on his screen.[10]

In 2005, former NASA flight director Gene Kranz spoke about the mishap that threatened the Apollo 12 mission and the actions within Mission Control to restore it to flight capability:

> We had three crewmen inside a dead spacecraft hurtling up towards orbit. The only thing working for us was the booster guidance system . . . Mission Control had a two-minute window of opportunity to save that mission. A controller makes a call to restore the instrumentation back onboard the spacecraft. We start voicing instructions up to verify reactant flow back to the fuel cells, bring them back on line. We needed to secure the navigation system in that two-minute window of opportunity. We may by now have saved the Apollo 12 mission. We got the crew up into orbit, checked them out, delayed injection to the Moon by one revolution, and in a gutsy move decided to send them to the Moon.[11]

Once crew and ground checks proved that there was no reason for the flight to be abandoned, the crew of Pete Conrad, Alan Bean and CMP Dick Gordon could concentrate on achieving orbit and then proceeding to their lunar destination: the area known as the Ocean of Storms.

On 19 November, Conrad and Bean made a successful, precision landing on the northwestern rim of Surveyor Crater. In fact, their touchdown had been so precise that they were located just 182 m (600 ft) from the unmanned probe Surveyor 3, which had landed on the Moon more than two years earlier, on 20 April 1967.

A renowned joker, the 5 ft 6 in. Pete Conrad was a little more light-hearted than Neil Armstrong when he stepped onto the Moon, with a joyful cry of 'Whoopie! Man, that may have been a small one for Neil, but that's a long one for me.' He then proceeded to hop and skip across the lunar surface. After joining him on the Moon, Alan Bean began to set up the colour television camera to record their activities, but as he went to place it on the tripod he accidentally allowed the camera to point directly at the Sun, instantly burning out the camera's sensor. TV viewers across the planet saw nothing more of the moonwalk than a blank screen.

During their first lunar EVA, of 3 hours and 56 minutes, Conrad and Bean planted the U.S. flag, collected some lunar rocks and deployed the first Apollo Lunar Surface Experiments Package (ALSEP) array of instruments. On their second EVA, the following day (3 hours, 49 minutes), the two moonwalkers paid a visit to the Surveyor craft, removing some parts for later analysis on Earth.

After spending 31 hours and 31 minutes on the Moon, lunar module *Intrepid* blasted off amid a shower of lunar dirt and dust, heading for a rendezvous with Dick Gordon aboard CSM *Yankee Clipper*. This took

Apollo 12 mission commander Pete Conrad studies the Surveyor 3 spacecraft. Lunar Module *Intrepid* can be seen on the lunar horizon.

place 3 hours and 30 minutes after lift-off, and once the two spacecraft had been reunited, Conrad and Bean went to move straight back into *Yankee Clipper*. But a horrified Dick Gordon would have none of that. He saw that their space suits were filthy with Moon dust and was aware that abrasive particles could float into and damage delicate instrumentation. 'You're not coming in my ship like that, Pete!' he yelled. 'Strip down!' Conrad and Bean looked at each other, shrugged, and took off literally everything. When they finally floated into the command module, they were both laughing and completely naked.

Once the fun was over and all the suits, samples and equipment had been transferred from *Intrepid*, the ascent module was detached, later crashing into the Moon, where the ALSEP's seismometer registered reverberations from the impact and transmitted this data back to Mission Control. It, too, has never been located.

Having orbited the Moon 45 times, and now with a full crew complement on board, *Yankee Clipper*'s engine fired and the crew was heading home, on the way witnessing a spectacular solar eclipse. Splashdown in the Pacific occurred on 24 November near the prime recovery vessel, aircraft carrier USS *Hornet*. Then, as with the crew of Apollo 11, Conrad, Bean and Gordon had to transfer into the Apollo quarantine caravan for three weeks, just to ensure they had not brought back any harmful bacterial bugs.[12]

### Survival in Space

Over four harrowing days filled with extreme tension in April 1970, people all around the world held their breath as Apollo 13 astronauts Jim Lovell, Fred Haise and Jack Swigert fought a desperate struggle for survival in space and to make it back home before their spacecraft became their coffin. Just 55 hours after being launched on NASA's third lunar landing expedition, an explosion in the cryogenic oxygen system blew the side out of their service module, causing NASA to abort the Moon landing and work frantically on unprecedented ways of getting the crew back to Earth before their supply of oxygen ran out. Though a catastrophic failure prevented the Apollo 13 mission from successfully achieving its objective, it became an epic story of survival against the odds and was later turned into a feature film.

A major problem occurred just two days before the scheduled launch, when CMP Ken Mattingly had to be replaced on the crew. Two

The Apollo 13 crew that would launch to the Moon: Commander Jim Lovell, Jack Swigert and Fred Haise.

weeks earlier, the crew and their backups had spent a last weekend relaxing with their families, but it was later revealed that one of the children had contracted measles. All of the crew members were cleared, having had the disease as children, but tests revealed that Mattingly – who was unsure if he had or not – was the exception. Even though it was a remote possibility, NASA did not want to send one of its crew on a ten-day mission during which he might develop the disease. After much discussion it was decided to replace Mattingly with his fully trained backup CMP, Jack Swigert. At the time it was a bitter pill for Mattingly to swallow, although he would later be reassigned to the Apollo 16 mission.

On 11 April 1970, Apollo 13 was launched on the planned ten-day mission to the Moon and back, with the objective of landing mission commander Jim Lovell and LMP Fred Haise in the Moon's Fra Mauro highlands. Even as their Saturn V rocket was sending them skyward, there were problems for the continuation of the mission, as outlined by flight director Gene Kranz:

> The mission is run on trust. You turn seven and a half million pounds of thrust loose, that is commitment! There is no changing your mind, and no turning back. Trust allows us to make every second count, and exploit every possible opportunity. We lost an engine during the second stage of powered flight. Very quickly, we looked at the remaining four engines – they were all go. We computed the new engines shutdown time, passed them up to the crew. Everything kept working, we got up into orbit. We checked the spacecraft out in orbit, and two and a half hours later we gave the go to go to the Moon.[13]

The S-IVB stage ignited as planned, transposition and docking with the LM was carried out flawlessly, and the crew of Apollo 13 were now on their way to a lunar rendezvous. Then, on the evening of 13 April, 55 hours into their outbound flight, all hell suddenly broke loose when a failure in the service module's cryogenic oxygen system led to a crippling explosion, which tore out the side of the module. Precious oxygen began venting out into space.

It all began when fellow astronaut and CapCom Jack Lousma reminded Swigert to carry out what was called a routine 'cryo-stir', which entailed flipping a switch on his console. This stir would allow information to be relayed on how much oxygen was left within a tank located in the service module. As director of flight crew operations Deke Slayton later explained: 'We later figured there were bare wires inside the tank – its insulation had been cooked off by mistake during some ground tests weeks before launch. The cryo stir caused a spark that started a fire inside the tank. When the pressure was too much, it just blew . . . taking the side of the service module with it.'[14]

At first the crew thought a meteoroid or orbiting space junk may have hit the spacecraft, but then it became apparent they were rapidly losing oxygen. 'It looks to me . . . that we are venting something,' Swigert said, peering out of CM *Odyssey*. 'It's a gas of some sort.' As red lights began appearing on the Emergency, Environmental, and Consumables Management (EECOM) console within Mission Control, the voice of Jim Lovell came through, dramatically saying, 'Houston, we've had a problem.' Which meant an instant end to any plans to land on the Moon. As Slayton later said: 'Pretty soon the fuel cells powering *Odyssey* began to fail. Oxygen pressure began to fall. No matter what Gene Kranz, the flight director, and his troops tried to do, nothing seemed to help. *Odyssey* was dying.'[15]

Even though the crew remained calm, they also knew they were in serious trouble as Mission Control worked frantically to come up with life-saving answers, as recalled by Gene Kranz:

> By the end of the second day, this crew is two hundred thousand miles from Earth. They are fifty thousand miles from the surface of the Moon, and entering a phase of the mission called entering the lunar sphere of influence. This is where we cross over from the boundary of the Earth's gravity to the lunar gravity. For a very short period of time – about four hours – you have two mission abort options. One goes around the front side of the Moon, the other goes completely around the Moon. But you have got to make up your mind quick, because for these options, time is running out.[16]

Fortunately, they were on their way to the Moon and not returning home after the landing, as they were still attached to LM *Aquarius*. Once the initial panic died down, everyone knuckled down to some tough and untried work. Although growing increasingly tired and cold, following the advice of Mission Control the crew performed a tricky engine burn to swing them around the far side of the Moon and head back to Earth.

Without electrical power, heating, computers or the use of their propulsion system, Lovell, Haise and Swigert were told they would have to evacuate into *Aquarius*, using it as an unintended lifeboat, while trying to find some way to purge their cramped living space of a potentially lethal build-up of carbon dioxide. They were tasked with solving this problem through the innovative use of objects carried within the spacecraft.

The crew knew that with every breath the three of them were overloading the carbon dioxide 'scrubbers' in the LM. These were basically filters filled with lithium hydroxide, designed to absorb the carbon dioxide. While there were several scrubbers located in the evacuated CM they were a different size and shape to those in *Aquarius*; the LM used cylindrical scrubbers while those in *Odyssey* were cubic. It was a classic case of trying to fit a square peg into a round hole. NASA's engineers got stuck into finding a workable solution using only the equipment on board and relaying the information verbally to the crew. Gradually, using hoses from spacesuits, duct tape and even their socks, the astronauts managed to fashion a square scrubber to fit the circular opening of the LM's filtration system. Then, exhausted, with very little sleep and working in freezing-cold conditions, they had to find ways of conserving what little electrical power they could muster and perform two course-correction manoeuvres, which they carried out perfectly.

It was a race against time, but on 16 April the crew moved back inside CM *Odyssey*, powered up, and discarded their service module.

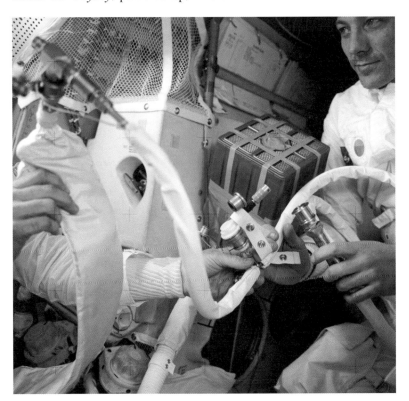

The Apollo 13 crew had to improvise and construct life-saving equipment using any available onboard materials.

For the first time they could see and photograph the full extent of the damage that had robbed them of a lunar landing. Then it was also time to jettison *Aquarius* (which had undoubtedly saved their lives) and prepare for re-entry. 'Farewell, *Aquarius*, and we thank you,' a grateful Lovell said as the crew watched their lunar module lifeboat slowly drift away, soon to be burnt up in its own re-entry.

With prayers that the crew had survived a fiery passage through the atmosphere, and following a period of communication blackout, CapCom Joe Kerwin kept calling *Odyssey*, but there was no response. Finally, to the relief and cheers of everyone, the calm voice of Jack Swigert came through: 'Okay, Joe.' It had been an extremely close-run thing. Splashdown came a little under six days into the troubled flight and ended within sight of the aircraft carrier uss *Iwo Jima*. Exhausted but exultant, the three crew members stood on the deck of the carrier, knowing they had come within a whisker of becoming space casualties and possibly terminating the Apollo lunar landing programme.

### Losing the Moon

The Space Race had been run and won with the successful Apollo 11 mission. America had beaten Russia to the Moon, but the euphoria of conquest had rapidly abated. As more astronauts left lingering foot-prints in the lunar surface, Americans – once enthralled by the great adventure – began to grow increasingly apathetic with the continuing Apollo programme. Some even took up the issue with their congress-men, demanding to know why NASA was spending even more millions gathering up a few extra Moon rocks when there were pressing social problems at home. The sight of astronauts gleefully bouncing around in one-sixth gravity had become tiresome for many, and public support for the space agency's lunar missions had plummeted. The dramatic life-or-death flight of Apollo 13 had amply demonstrated the colossal risks associated with human-tended space missions, but once the crew was safely back home, a widespread apathy for further space exploration set in again.

In January 1970, prior to the Apollo 13 mission, and with NASA now facing severe budgetary woes, the space agency reluctantly decided to pare Apollo 20 from the flight manifest. The late President Kennedy had been right: landing astronauts on another world was a monu-mentally costly exercise. Eight months later, on 2 September 1970, the budgetary axe fell once again and NASA lost yet another two lunar missions. The Apollo 18 and 19 missions were cancelled, and in a re-shuffle of mission objectives the final three flights were renumbered as Apollo 15, 16 and 17.

Nevertheless, the lunar programme would continue four months after the loss of the last three Apollo missions. Apollo 14 launched to the Moon on 31 January 1971 and became the third mission to achieve a lunar landing. Commander Alan Shepard and LMP Edgar Mitchell touched down in the Fra Mauro highlands, earlier the target area for the crew of Apollo 13. As described by the Lunar and Planetary Institute in Houston:

> The landing site is located in a broad, shallow valley between radial ridges of the Fra Mauro Formation and approximately 500 kilometers from the edge of the Imbrium Basin. The major crater Copernicus lies 360 kilometers to the north, and bright ray material that emanates from Copernicus Crater covers much of the landing site region. In the immediate landing site area, an important feature is the young, very blocky Cone Crater, which is approximately 340 meters in diameter and which penetrates the regolith on the ridge to the east of the landing site.[17]

For Alan Shepard, the command of Apollo 14 fulfilled a long-held desire to fly into space one more time. A decade before, he had become the first American to fly into space, but that space flight experience had lasted only fifteen minutes. Following Gordon Cooper's Mercury mission aboard *Faith 7* he had tried everything he knew and every important contact he had – right up to President Kennedy – to try and secure one last long-duration Mercury space flight. After all, he argued, there was a spare Mercury capsule (Spacecraft 15B, the backup vehicle for Cooper's mission) available. But the president diplomatically referred Shepard back to NASA administrator James Webb, who was adamant that Project Mercury was over, and most of the agency's personnel were now busy working on Project Gemini.

Then came Shepard's problem with the dizzying inner-ear problem Ménière's disease, which caused him to lose flight status and his place on the first Gemini mission. He feared his astronaut career was at an end and glumly accepted the interim role of chief of the Astronaut Office. In 1968 he underwent a risky but successful ear operation, and

Alan Shepard returned to space as commander of the Apollo 14 mission. From left: Stuart Roosa, Shepard and Edgar Mitchell.

he regained his active astronaut status. Desperate to command an Apollo mission, he and Deke Slayton had discussed the possibility of Shepard taking over command of the tasks ahead of him from Gordon Cooper, who, as backup commander for Apollo 10, was in line to rotate to Apollo 13. However, Cooper was being perceived as becoming increasingly rebellious and lax in his Apollo training. Shepard, knowing this, had no qualms about bouncing his Mercury colleague and stepped into the command role. Cooper, appalled, quit NASA in disgust. The next problem was a lack of sufficient training for Shepard and his crew, which was resolved by switching Shepard's crew to Apollo 14, and Jim Lovell's well-trained crew taking on Apollo 13.

Now the Apollo 14 crew was settled: Alan Shepard as commander, Stuart Roosa as CMP and Edgar Mitchell as LMP. Neither of Shepard's crewmates had previous missions under their belts, so when Apollo 14 flew to the Moon, their collective space flight experience was a mere fifteen minutes.

On this flight, the crew had the added facility on the Moon's surface of a collapsible, two-wheeled cart known as the Modular Equipment Transporter (MET), allowing the astronauts to not only travel further from the LM than previous crews but to load up the MET and take several tools, cameras and a portable magnetometer with them.

During their 33.5 hours on the Moon, Shepard and Mitchell carried out two lunar EVAs, totalling over nine hours on the lunar surface, covering just over 3.5 km (2 mi.) from the lunar module, *Antares*, and travelling to thirteen locations, setting up ten experiments while exploring and photographing areas and features of peculiar interest. They had collected some 43 kg (94 lb) of rocks and soil for later study and analysis back on Earth.

On their second EVA, Shepard and Mitchell had attempted to reach the rim of Cone Crater, approximately 90 m (300 ft) above their landing site. They were moving in on the crater when Mission Control warned them they had run out of time for the venture and to gather up some samples from where they stood and return to *Antares*. On this EVA, Shepard set a new record for the longest distance travelled on the Moon by an astronaut, of approximately 2.74 km (1.7 mi.).

Ask anyone, however, and Apollo 14 is usually most remembered for Alan Shepard constructing a modified golf stick out of a digging tool shaft and a club head he had brought with him and belting two golf balls one-handed across the surface of the Moon. Shepard would laughingly say after the mission that the balls had travelled 'miles and miles', but in reality he could put little energy into a one-handed swing in a pressurized space suit, and the balls travelled only a few metres.

*Antares* lifted off from the Moon on 6 February and successfully docked with CSM *Kittyhawk*, flown by Stuart Roosa. The mission ended with a safe splashdown on 9 February 1971. This would be the first crew that did not have to suffer the ignominy of being held for a time in the portable quarantine facility once it had been proven that the Moon is sterile. The quarantine period had been eliminated, much to the relief of this and future Apollo crews.

### Driving into History

Apollo 15 was the first of three missions designed to conduct far more extensive exploration of the Moon over much longer periods, which would be facilitated by the installation on the lunar module of the first Lunar Roving Vehicle (LRV). Following a flawless on-schedule lift-off from the Kennedy Space Center's Launch Complex 39 at 9:34 a.m. EDT on 26 July 1971, the crew encountered very few problems as they headed for a landing near the foot of the Montes Apenninus (Appennine Mountains) and adjacent to Hadley Rille, a geologically rich area to explore. Having achieved lunar orbit, commander Dave Scott and LMP Jim Irwin transferred into LM *Falcon*, leaving Al Worden to fly solo in CSM *Endeavour*. A landing was completed on a dark plain near Hadley Rille. Then, during three periods of lunar EVA over a three-day period, Scott and Irwin completed a record 18 hours and 37 minutes exploring the lunar surface and various features.

The LRV proved a great success, and the two astronauts were able to travel over 27 km (17 mi.) in the vehicle, collecting more than 77 kg (170 lb) of lunar samples, obtain a core sample collected from 3 m (10 ft) beneath the lunar surface, set up instruments and experiments, and take hundreds of photographs.

On 2 August, *Falcon*'s single engine was fired, lifting the ascent stage engine from the now-abandoned descent stage. A successful rendezvous and docking was then completed with CSM *Endeavour* on its fiftieth lunar orbit. On the 74th orbit, the crew launched a small Particles and Fields Subsatellite (PFS) from the service module. On the very next revolution, a lengthy engine burn of 2 minutes and 21 seconds began *Endeavour*'s flight back to Earth.

The crew of Apollo 15: Commander Dave Scott, CMP Al Worden and LMP Jim Irwin.

Irwin salutes the U.S. flag in front of LM *Falcon*.

On 5 August, Al Worden became the first human to carry out a deep-space operational EVA, exiting the CM at the end of a tether and carefully climbing to the rear of the service module, where he retrieved some film cassettes from the scientific instrument module (SIM) bay cameras before returning to the hatch, where Jim Irwin was positioned halfway out to receive the cassettes. Like Dave Scott inside the spacecraft's CM, he was wearing his full protective space suit. As he exited the hatch, Worden heard from fellow astronaut Karl Henize in Mission Control telling him there was no reason to hurry back inside *Endeavour*, but Worden was fully focused on the tasks he had to perform. It was only later, back inside *Endeavour* and with the hatch closed, that he realized he should have taken Henize's advice and taken time to absorb the awesome splendour that surrounded them. He had, however, taken a quick look around while planting his boots into special restraints and waiting for Jim to float into position. His thoughts at that time are best captured in Worden's own words from his later memoir:

> I hadn't really had a sense of where I was until this moment. Standing upright on the side of the spacecraft, attached only by my feet and the umbilical that loosely snaked back to the spacecraft hatch, I had a fleeting sense of being deep under the ocean, in the dark, next to an enormous white whale. The sun was at a low angle behind me, so every bump on the outside of the service module cast a deep shadow. I didn't dare look toward the sun, knowing it would be blindingly bright. In the other direction, and all around me, there was – *nothing*. It's a sensation impossible to experience unless you float tens of thousands of miles from the nearest planet. This wasn't deep, dark water, or night sky, or any other wide open space that I could comprehend. The blackness defied understanding, because it stretched away from me for billions of miles.[18]

It was with a deep regret that Worden finally unhooked his boots from the restraints and turned to make his way back inside at the end of his historic eighteen-minute excursion, but there was one more astonishing sight that made him gasp in amazement. He saw Jim Irwin, halfway out of the hatch, framed by an enormous silver Moon right behind him. He later said that if he had taken a camera out with him, it would have been one of the most famous photos ever taken in the space programme.

Two days later, Apollo 15 splashed down in the Pacific Ocean north of Honolulu, ending a flight of 12 days and 7 hours. Soon after, the crew was carried by helicopters to the prime recovery ship, USS *Okinawa*, close to the targeted touchdown point.

### Exploring Descartes

Apollo 16, the penultimate crewed mission to the Moon and the second of three extensive lunar exploration missions, began with a near-perfect lift-off from Launch Pad 39A at the Kennedy Space Center on 16 April 1972. All went well until lunar module *Orion* was powered up in lunar orbit with commander John Young and LMP Charlie Duke aboard, ready to begin their descent to the Moon, leaving Ken Mattingly alone in CM *Casper*.

To prepare for the possibility of an aborted landing, Mattingly would have to practise guiding *Casper* to a rendezvous and docking in lunar orbit with *Orion*, requiring a test burn of the CM's steerable rocket engine. During the test, a malfunction was detected which Mattingly had to attempt to restore before the Moon landing could take place. After six hours, during which time Young and Duke could only patiently bide their time aboard *Orion*, NASA decided that tests demonstrated they could work around the problem if need be, and Young, to his relief, was informed the landing could proceed.

Young and Duke skilfully piloted *Orion* to a touchdown on the western edge of the Descartes Mountains in the central lunar highlands on 20 April. During their 71 hours and 2 minutes on the surface of the Moon, the two astronauts carried out three lunar EVAs totalling 20 hours and 14 minutes with the aid of their lunar rover, driving for just over 26 km (16 mi.) and gathering together 96.6 kg (218 lb) of Moon rocks

The crew assigned to the Apollo 16 mission: CMP Ken Mattingly, mission commander John Young and LMP Charlie Duke.

A video camera mounted on the LRV showed the lift-off of *Orion*'s ascent stage.

and other samples, as well as setting up instruments, conducting tests and drilling core samples. In order to maintain a strict timetable, the third EVA, at 5 hours and 49 minutes, was around two hours shorter than the previous two excursions as a result of the late landing. A highlight for them was locating one of the largest boulders ever seen on the Moon, which they dubbed 'House Rock'. At the conclusion of this third lunar excursion, Young dropped Duke near *Orion* and then took the LRV for a solo spin in which he reached a top speed of 18 km/h (11 mph), kicking up a lot of dust before having to abandon the vehicle and prepare to leave Descartes.

A camera mounted on the LRV, directed from Mission Control in Houston, was able to record *Orion*'s ascent engine firing and track the craft as it roared upward to an eventual rendezvous and docking with CM *Casper*. After 64 orbits of the Moon, launching a subsatellite and disengaging from the now vacated *Orion*, *Casper*'s engine fired and the Apollo 16 crew were heading for home. During the return journey, Ken Mattingly performed a similar deep-space EVA to that of Al Worden on the previous mission, spending 1 hour and 24 minutes outside while retrieving film cassettes from the SIM bay. The crew splashed down safely just 1.5 km (1 mi.) from their prime recovery vessel, the aircraft carrier USS *Ticonderoga*, and were on board just 37 minutes later.[19]

### Last Man on the Moon

The stainless steel plaque was attached to the ladder on the descent stage of LM *Challenger*, and it read: 'Here man completed his first exploration of the Moon, December, 1972. May the spirit of peace in which we came be reflected in the lives of all mankind.' The final Apollo excursion to the lunar surface took place in the Taurus–Littrow valley, named for the surrounding Taurus Mountains and the massive nearby Littrow crater. The site was chosen as it promised a rich variety of collectable samples dating from ancient highland and much younger areas of volcanic activity. This final lunar mission was under the command of Captain Eugene 'Gene' Cernan, accompanied by geologist Harrison ('Jack') Schmitt. Once they had completed the last of their three lunar excursions, and with all the specimen material they had gathered and other equipment safely stowed aboard *Challenger*, it was time for a few moments of reflection.

As Cernan prepared to step off the Moon, he left behind in the lunar soil the last footprint of the first generation that had challenged the new frontier of space, but with no known plans for a return to the Moon. An eloquent man, Cernan paused and observed, 'As I take these last steps from the surface for some time into the future to come, I'd just like to record that America's challenge of today has forged man's destiny of tomorrow.' He added, 'And as we leave the Moon at Taurus–Littrow we leave as we came, and, God willing, we shall return, with peace and hope for all mankind.' He then followed Jack Schmitt up the skinny ladder into the cramped cabin of *Challenger*.[20] He was not to know it then, but at the time of his death in 2017, Gene Cernan still held a place in the history books as the last man to set foot on the Moon.

Cernan and Schmitt, the eleventh and twelfth men to visit the Moon, had spent more time on the lunar surface than any previous Apollo mission – 22 hours, 5 minutes – and covered the greatest distance – 35.4 km (22 mi.) – during their three EVA excursions. It had been a remarkably ambitious and successful mission with which to end Project Apollo's lunar landing programme.

It all began after a delay of 2 hours and 40 minutes resulting from a computer glitch, when the launch of Apollo 17 took place from Kennedy Space Center's Launch Pad 39A at 12:33 a.m. on 7 December 1972. Owing to orbital mechanics, it marked the first night-time launch of a gigantic Saturn v rocket. On board were mission commander Gene Cernan and CMP Ron Evans, together with scientist-astronaut and LMP Harrison 'Jack' Schmitt. Schmitt had been the subject of a change to the original crew, taking over the LMP spot originally allocated to astronaut and former x-15 pilot Joe Engle. With the cancellation of the final three Apollo missions, 18, 19 and 20, Schmitt, a trained geologist, was suddenly without a flight to the Moon, where his geological background would have proved invaluable. Pressure was subsequently exerted on NASA to place Schmitt on the final crew, and a disappointed Engle lost his place as LMP on what would be the final lunar landing mission.[21] At the moment of lift-off, the Saturn rocket's five massive F1 engines howled into life in controlled but violent fury, turning Florida's midnight sky into a blazing dawn.

Once they were safely in orbit, the crew prepared for the moment their s-IVB third-stage rocket would ignite, with the welcome command to proceed: 'Translunar injection.' When that confirmation came through, the reliable engine on the third stage began firing once again, pushing the astronauts back into their couches and accelerating their spacecraft to around 40,000 km/h (25,000 mph), now heading out of Earth orbit for a rendezvous with the Moon.

The crew of the final lunar landing mission, Apollo 17: at rear, LMP Harrison (Jack) Schmitt and CMP Ron Evans, with mission commander Gene Cernan at front.

## Apollo Crewed Missions, 1968–72

| Apollo Mission | Crew (CDR/CMP/LMP) | Launch and Landing Dates | Mission Highlights |
|---|---|---|---|
| Apollo 7 (AS-7) | Walter M. Schirra Jr<br>Donn F. Eisele<br>R. Walter Cunningham | 11 October 1968<br>22 October 1969 | First crewed Apollo flight after Apollo 1 fire |
| Apollo 8 (AS-8) | Frank F. Borman II<br>James A. Lovell Jr<br>William A. Anders | 21 December 1968<br>27 December 1968 | First crewed flight around the Moon |
| Apollo 9 (AS-9) | James A. McDivitt<br>David R. Scott<br>Russell L. Schweickart | 3 March 1969<br>13 March 1969 | First test of lunar module in Earth orbit |
| Apollo 10 (AS-10) | Thomas P. Stafford<br>John W. Young<br>Eugene A. Cernan | 18 May 1969<br>26 May 1969 | Successful rehearsal of first Moon landing |
| Apollo 11 (AS-11) | Neil A. Armstrong<br>Michael Collins<br>Edward E. 'Buzz' Aldrin | 16 July 1969<br>24 July 1969 | First humans to walk on the Moon's surface |
| Apollo 12 (AS-12) | Charles 'Pete' Conrad Jr<br>Richard F. Gordon Jr<br>Alan L. Bean | 14 November 1969<br>24 November 1969 | Second successful landing on the Moon |
| Apollo 13 (AS-11) | James A. Lovell Jr<br>John L. 'Jack' Swigert Jr<br>Fred W. Haise Jr | 11 April 1970<br>17 April 1970 | Mission abandoned following explosion in CSM |
| Apollo 14 (AS-14) | Alan B. Shepard Jr<br>Stuart A. Roosa<br>Edgar D. Mitchell | 31 January 1971<br>9 February 1971 | Longest lunar excursion and first golf shot on the Moon |
| Apollo 15 (AS-15) | David R. Scott<br>Alfred M. Worden<br>James B. Irwin | 26 July 1971<br>7 August 1971 | First use of Lunar Rover Vehicle (LRV) on Moon |
| Apollo 16 (AS-16) | John W. Young<br>Thomas K. Mattingly II<br>Charles M. Duke Jr | 16 April 1972<br>27 April 1972 | Record time spent exploring lunar surface with LRV |
| Apollo 17 (AS-17) | Eugene A. Cernan<br>Ronald E. Evans Jr<br>Harrison H. 'Jack' Schmitt | 7 December 1972<br>19 December 1972 | Final Apollo mission and last person on the Moon |

Their three-day outward journey was busy but smooth and un-eventful, and as they entered into lunar orbit the three astronauts began making preparations for the following day's landing, on a mountain-fringed site in the Moon's Taurus–Littrow region. Their prime target was a crater 610 m (2,000 ft) in circumference, believed to be filled with ash from an ancient volcano, which could provide scientists with both the oldest and youngest lunar material ever seen. High-resolution images taken of the area by a previous Apollo crew had identified the crater as

This photograph of the full Earth, a shot famously known as 'The Blue Marble,' is believed to have been taken by LMP Jack Schmitt on their outward journey, although NASA credits the photo to the entire crew.

being of particular interest, and during their training Cernan and Schmitt had named it Camelot, in honour of the late President Kennedy's administration.

All went well with the separation from CSM *America*, now in the solo hands of Ron Evans, and the landing was so perfect that LM *Challenger* had extra fuel left in its descent stage after the twelve-minute power dive from lunar orbit. Once he had clambered down the LM ladder and had a chance to look around, Cernan compared the mile-high hills to the east of the touchdown site to the 'skin texture of an old, old, hundred-year-old man'. Schmitt was also fascinated by the stark scene around him, telling Houston that it was 'a different breed of rock up here' to those found on the five previous lunar expeditions.

After planting the U.S. flag at the landing site, Cernan and Schmitt set off to explore the area in their electric-powered lunar rover on the longest and most complex trio of lunar excursions to that time, although they had to make temporary repairs to a broken fender on the

LRV which had caused sticky Moon dust to settle everywhere in and on the vehicle.

On their furthest excursion from the landing site they drove 8 km (5 mi.) from *Challenger* as they sought out unique places and formations to sample. One of these was a patch of orange soil, which Schmitt thought might have indicated recent volcanic activity and possible water on the Moon. Schmitt also picked up a Moon rock which he reported was a fused mixture of 'fragments of all sizes and shapes, and even colours that had grown together and . . . sort of living together in a very peaceful manner'. Another task was to erect a nuclear-powered science station, which promptly began sending data back to Earth.[22]

Their work finally at an end after a record 75-hour visit, Cernan and Schmitt prepared to lift off from the Moon. As before, a colour

Jack Schmitt by the massive Tracy's Rock, located at the EVA's Station 6.

television camera mounted on the LRV sent back footage of *Challenger* as it lifted off and soared into the darkness in a shower of debris, hurtling upward, ready to rendezvous with Ron Evans, who had spent three days in solo lunar orbit aboard *America*.

On the second attempt a hard dock was achieved, which enabled Cernan and Schmitt to make their way into *America* through a connecting tunnel and to a joyous reunion with Evans. They later transferred 101.6 kg (224 lb) of samples into *America* for the journey home, where scientists were eagerly awaiting the rocks and soil and the return of trained geologist Schmitt. They would have to curb their impatience for a while, as the crew remained in lunar orbit for nearly two more days after discarding LM *Challenger* – which later crashed onto the Moon. The crew spent this time gathering close-up lunar data and making visual observations, knowing they might be the last lunar visitors for some time to come.

On their return journey, Ron Evans completed the customary deep-space EVA (this time lasting 66 minutes) to retrieve film packages from the SIM bay. Twelve and a half days after lifting off from the Kennedy Space Center, *America* splashed down close to the recovery ship, USS *Ticonderoga*. Once the flotation collar had been fitted by frogmen, the three astronauts clambered out and were whisked across to the aircraft carrier for medical tests and a debriefing.

Project Apollo's lunar landing programme was at an end, but the six successful missions had brought back a bounteous haul of material. According to NASA, 'Between 1969 and 1972 six Apollo missions brought back 382 kilograms (842 pounds) of lunar rocks, core samples, pebbles, sand and dust from the lunar surface. The six space flights returned 2,200 separate samples from six different exploration sites on the Moon.'[23]

# 8

# SOVIET SETBACK AND SKYLAB

With the impending termination of the lunar landing programme after Apollo 17, some of NASA's focus was already beginning to turn to the creation of crewed orbital workshops, using Saturn rockets and hardware that would have been launched on the three cancelled Apollo lunar landing missions. Meanwhile, the Soviet Union, having abandoned plans to fly cosmonauts to the Moon following the loss of four of its mighty N-1 rockets during launch, resumed their Soyuz programme.

Eighteen months after the death of Vladimir Komarov on the ill-fated Soyuz-1 mission, Soyuz-2 was launched on a crewless flight from the Baikonur Cosmodrome. It was followed into orbit three days later by Soyuz-3, piloted by Georgi Beregovoi, with plans for the cosmonaut to dock with the unmanned craft. Problems continued to plague the Soyuz programme when Beregovoi tried but failed to dock manually with Soyuz-2, continually overriding the automatic systems in his attempts. In doing so he expended most of his orientation fuel and as a result was forced to abandon his efforts and return to Earth. Incredibly, he had not realized that the Soyuz-2 vehicle was actually upside down in relation to his spacecraft. The Soviet media tried to downplay this failure by stating that all the mission objectives had been met, but Beregovoi would never fly into space again.

In January 1969, with the Apollo programme now in full swing ahead of the first crewed landing, Vladimir Shatalov was sent aloft aboard Soyuz-4, joined in orbit the next day by the three-man crew of Soyuz-5. This time, on 16 January, the docking was successful – the first time two crewed spacecraft from either nation had linked up in orbit. While they were joined together, two of the Soyuz-5 crew, Yevgeny Khrunov and flight engineer Alexei Yeliseyev, donned Yastreb (Hawk) spacesuits and conducted an EVA to the other spacecraft, leaving Boris Volynov alone in Soyuz-5 and joining Shatalov aboard Soyuz-4. After the crew transfer and undocking, the two Soyuz craft separated. Soyuz-4 re-entered the following day, the flight ending with a successful touchdown. Volynov, now flying solo, would land the day after.[1]

Soyuz-6 was launched into orbit nine months later, followed the next day by Soyuz-7, and the day after that Soyuz-8 joined the other two craft in orbit as part of a joint mission. Three spacecraft were now in orbit together for the first time, carrying a total of seven cosmonauts. The major objective of the triple mission was for Soyuz-7 to dock with Soyuz-8, while the nearby crew of Soyuz-6 filmed the entire operation. Much to everyone's chagrin, the docking was unsuccessful (although TASS once again reported that this manoeuvre had never been intended for the flight), and one by one the three Soyuz craft returned to a safe landing.

Although any interest in crewed space flights had rapidly collapsed in certain quarters of Soviet leadership, powerful General Secretary Leonid Brezhnev declared that a programme of Earth-orbiting space stations should be immediately undertaken, particularly in light of America's interest in developing a similar programme. Brezhnev therefore ordered a cooperative crash programme within all of the Soviet

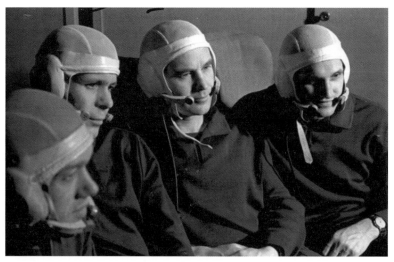

The four cosmonauts involved in the Soyuz-4 and Soyuz-5 docking mission. From left: Alexei Yeliseyev, Yevgeny Khrunov, Vladimir Shatalov and Boris Volynov.

## The First Soviet Soyuz Missions

| Flight | Crew | Launch | Landing | Mission Objectives/Results |
|--------|------|--------|---------|----------------------------|
| Soyuz-1 | Vladimir Komarov | 23 April 1967 | 24 April 1967 | Mission aborted: cosmonaut killed after parachute failure |
| Soyuz-2 | None | 25 October 1968 | 28 October 1968 | Test flight as intended docking target for Soyuz-3 |
| Soyuz-3 | Georgi Beregovoi | 28 October 1968 | 30 October 1968 | Failed docking with Soyuz-2 |
| Soyuz-4 | Vladimir Shatalov | 14 January 1969 | 17 January 1969 | Docking with Soyuz-5, two-crew transfer via EVA |
| Soyuz-5 | Alexei Yeliseyev Yevgeny Khrunov Boris Volynov | 15 January 1969 | 18 January 1969 | Yeliseyev and Khrunov transfer to Soyuz-4. Volynov lands alone |
| Soyuz-6 | Georgi Shonin Valery Kubasov | 11 October 1969 | 16 October 1969 | Joint orbital mission with Soyuz-7 and Soyuz-8 |
| Soyuz-7 | Anatoly Filipchenko Vladislav Volkov Viktor Gorbatko | 12 October 1969 | 17 October 1969 | Joint flight with Soyuz-6 and Soyuz-8 |
| Soyuz-8 | Vladimir Shatalov Alexei Yeliseyev | 13 October 1969 | 17 October 1969 | Joint flight with Soyuz-6 and Soyuz-7 |
| Soyuz-9 | Andrian Nikolayev Vitaly Sevastyanov | 1 June 1970 | 19 June 1970 | Prolonged endurance flight ahead of Salyut space station missions |

design bureaux to develop, construct and launch a civilian space station ahead of America's Skylab.

On 19 April 1971 the world's first space station, Salyut-1, also designated DOS-1 (Dolgovremennaya Orbitalnaya Stanziya, or Long Duration Orbital Station) was launched into low Earth orbit, circling the planet at an altitude ranging between 200 and 220 km (124–138 mi.). The TASS news agency reported that the space laboratory was circling the planet every 88.5 minutes, at an angle of 51.6 degrees to the equator. TASS said Salyut-1 was launched 'to perfect elements of the design and on-board systems and for conducting scientific research and experiments in space'.[2] It further stated that the station – whose name translates to 'Salute' – was functioning normally, but gave no immediate details of its size or whether it was capable of carrying men in orbit. Nevertheless, it was fully expected in the West that a manned launch to Salyut-1 was imminent.

Just four days later, in the early morning of 23 April, Soyuz-10 was launched into space on a mission to dock with Salyut-1. The three cosmonauts on board, Vladimir Shatalov, Alexei Yeliseyev and first-time civilian space traveller Nikolai Rukavishnikov, were able to successfully navigate their spacecraft to the orbiting station, and docking manoeuvres began when Salyut-1 was on its 86th orbit and Soyuz-10 its 12th. The crew's problems began when an automatic hard docking could not be achieved due to the angle of approach. Soyuz-10 was eventually docked manually, but after fifteen minutes Shatalov reported that he could not get a docking light, indicating a lack of electrical connection between the spacecraft and the station.

Ground telemetry showed that the two craft were still separated by a 9-centimetre (3½ in.) gap, making it impossible to obtain an airtight seal in the docking mechanisms. The frustrated crew, unable to board Salyut-1, could do little but sit and wait for further advice from the ground. Acting on instructions, they carried out all of ground control's recommended actions. Possible variants were attempted, and Shatalov tried repeatedly to force the docking, but nothing worked. After 5 hours and 30 minutes soft-docked to the station, the order was reluctantly given to terminate the attempt and withdraw.

Then, a further complication arose when the Soyuz probe could not be released from the station's docking cone despite Shatalov's best efforts. One solution would have been to jettison the orbital module on Soyuz-10 and leave it attached to Salyut-1, but this would make it impossible

The Soyuz-11 crew of Viktor Patsayev, Vladislav Volkov and Georgi Dobrovolsky relaxing during mission training.

for future Soyuz missions to dock, meaning the station would have to be abandoned. Eventually, ground controllers resolved the situation when they discovered that the crews could simply throw a circuit breaker in the docking mechanism, and this interruption to the power supply would cause the probe to automatically retract. To everyone's relief, this worked. The undocking was accomplished and the crew was ordered to return to the ground. Another problem occurred during re-entry when some toxic fumes began to fill the spacecraft, causing Rukavishnikov to pass out. Despite this, the landing was normal and all three crew members were recovered unscathed.[3]

## A Shock Loss

Initially, Georgi Dobrovolsky, Vladislav Volkov and Viktor Patsayev were assigned to serve as the backup crew for the Soyuz-11 mission to the Salyut-1 space station. The prime crew for the flight was meant to be Alexei Leonov, famed for his history-making spacewalk six years earlier, and Valery Kubasov, veteran of the Soyuz-6 mission, while Pyotr Kolodin would be making his maiden space flight. They had served as the backup crew for the troubled flight of Soyuz-10, forced to return prematurely following a docking failure with the orbiting station, and would take on the lengthy mission originally planned for the earlier crew.

Preparations for the flight continued to go well. Then, a week before the launch date, Kubasov underwent a routine medical check, which revealed what appeared to be a lung ailment. It was later discovered that this was nothing more than an allergic reaction to a pesticide commonly used at the Baikonur launch site. At the time, it was planned to drop Kubasov from the prime crew and replace him with aviation engineer Vladislav Volkov from the backup team. It was then determined that, coming so soon before the launch, this crew change might prove too disruptive, and a decision was instead made to replace the entire prime crew with the backup team. Leonov was furious with the decision and made his feelings known, but he was overruled. Dobrovolsky would now serve as mission commander alongside design engineer Viktor Patsayev, with Volkov remaining as part of the crew.

Despite the disruptions caused by a total crew change, the launch went ahead on 6 June 1971. At 4:55 a.m. the Soyuz-11 spacecraft left the Baikonur launch pad on a flawless trajectory and thundered into orbit. Unlike the previous failed mission, Soyuz-11 docked firmly with Salyut-1 the following day, enabling the crew to transfer into the orbiting laboratory – the first crew ever to enter a space station. The joined configuration of the two spacecraft measured 21.4 m (70 ft), and there was a total capacity of 100 cubic metres (3,530 cubic ft) – plenty of room for the cosmonauts to work, relax and sleep. News of the successful docking, along with the names of the crew, was then broadcast over Radio Moscow.

Two weeks into their mission, the three cosmonauts broke the space endurance record set five years earlier by NASA astronauts Frank Borman and Jim Lovell aboard Gemini 7, and still they flew on, later exceeding the near eighteen-day mission of the Soyuz-9 crew of Andrian Nikolayev and Vitaly Sevastyanov. The flight, however, was not without its problems. One serious episode occurred when the crew detected a small electrical fire in one of the station's systems, which caused the interior to fill with acrid smoke. Fortunately, they were able to extinguish the fire and resume their work schedule. Like other problems at that time, any reports of the incident were quickly denied and covered up by the Soviet space chiefs, only to become known several years later.

On 29 June, after nearly 24 days in orbit – and thereby doubling Gemini 7's endurance record – the crew re-entered their Soyuz-11 spacecraft. Dobrovolsky undocked the ship from Salyut-1 and prepared for their return to Earth. With three crew members aboard they would re-enter without wearing their protective space suits, creating more room and comfort for the cosmonauts, albeit creating a potentially hazardous situation.

After three additional orbits, Dobrovolsky notified ground control that they were ready to begin their descent. Soon after, as Soyuz-11 plunged ever deeper into the atmosphere, all contact was lost with the ground – a normal event that generally lasted around four minutes, as radio waves are unable to penetrate the heat and ionization associated with re-entry. Minutes later, when radio contact should have been restored with the descending spacecraft, there was nothing but silence, causing growing concern among the ground controllers, who repeatedly called the crew.

Meanwhile, on the steppes of Kazakhstan, everything appeared to be quite normal. Recovery helicopter crews scanned the skies and reported seeing the charred Soyuz descent module drifting down under its huge red-and-white parachute. When it approached the ground, braking rockets automatically fired, absorbing the shock of impact for

An illustration of the link-up between Soyuz-11 and Salyut-1.

the crew. As it settled and tilted, helicopters touched down nearby and would-be rescuers rushed over. Apart from the spacecraft lying on its side, which would make the crew extraction awkward, nothing seemed out of the ordinary according to the recovery teams. Ground controllers decided that the spacecraft's radio must have been faulty.

Amid great excitement, the rescuers banged on the side of the charred spacecraft, but there was no response. Instead, they encountered an eerie stillness and silence. When they opened the hatch, they found the three cosmonauts sitting motionless in their couches, with dark blue bruising on their faces and traces of blood coming from their ears and noses. The bodies were hastily removed through the hatch and unsuccessful attempts were made to revive the men using artificial respiration. The grim news was relayed to Moscow, where government officials had little choice but to go public with the terrible news.[4]

In addition to expressing horror at the deaths of the three cosmonauts, there was immediate concern in the West, especially when Soviet space officials remained tight-lipped as to the cause of the tragedy. It was feared that the mystery deaths could have resulted from a complete breakdown of the cosmonauts' vital organs, possibly made sluggish in protracted weightlessness and subsequently enduring the sudden stress of gravity.[5] The u.s. was demanding some swift and cooperative answers, as the fatalities could have serious implications for its own long-term space plans. NASA was looking at a series of protracted post-Apollo science missions aboard a Skylab orbiting space station, whose crews would easily eclipse the 23 days spent in orbit by the Soyuz-II crew. Twelve days later, the Soviets relented, but still withheld vital details. They would reveal only that the three cosmonauts had died of embolisms, or air bubbles in the blood, caused by sudden depressurization.[6]

It would be some three years before Soviet authorities finally disclosed the actual cause of the depressurization, traced to a small pressure equalization valve that had popped open during the separation of the descent module from the orbital and service modules. This valve had been designed to equalize air pressure between the cabin and the ambient outside air pressure once the spacecraft was descending through the denser layers of the atmosphere. Tragically, it had opened prematurely during the return trajectory, rapidly venting the cabin's atmosphere. Since none of the three cosmonauts were wearing pressure suits or helmets for launch and re-entry, they were totally unprotected from such a mishap. There is evidence to suggest that they realized there was a leak and tried to locate the source, but, being exposed to the vacuum of space, all three quickly fell unconscious and died of asphyxiation mere minutes away from the end of their record-breaking flight.[7] It was later suggested that the simple act of placing a finger over the leak might have saved their lives.

Dobrovolsky, Volkov and Patsayev received a solemn state funeral, with all three bodies dressed in civilian clothing. Following the service and cremation, the three men's ashes were inurned in Moscow's Kremlin Wall Necropolis, close to those of earlier casualties Yuri Gagarin and Vladimir Komarov. They were the third, fourth and fifth cosmonauts to have their ashes consigned there, and they would also be the last. With the later demise of the Soviet Union, burials in the wall came to an end.

The Salyut-I space station, originally designed to remain in orbit for just three months, would never again be occupied. Although boosted to a higher orbit in order to prolong its life, the station's reaction control system had become increasingly wayward following an electrical failure in October. The decision was made to de-orbit Salyut-I as soon as possible before complete control from the ground was lost and to bring it down safely over the Pacific Ocean. On 10 October 1971, the station's main engines were fired from the ground and, following

175 days in orbit, the world's first space station entered the atmosphere and was mostly incinerated before splashing down in the sea.

### Skylab: America's First Space Station

Originally known as the Apollo Applications Program, Skylab evolved from plans to use converted Apollo lunar hardware to create America's first space station. Once it had been placed in orbit, the Skylab laboratory would become a unique platform for research into solar science, stellar astronomy, space physics, Earth resources and life and material sciences. It was planned as a unique facility where visiting crew members would learn the skills and realities of living and working over extended periods in space.

Skylab became the largest spacecraft ever launched into orbit to that time. Fully equipped for human occupancy, it would be described in newspapers as 'a house in space'. From May 1973 to February 1974, three teams of astronauts occupied the station on dedicated science missions of 28, 59 and 84 days respectively. Skylab was not just a temporary workplace for American astronauts; it would become the forerunner to NASA's vision for the future, which the space agency called 'the next logical step in space' – creating a permanent human presence in Earth orbit. The entire programme, however, came perilously close to being abandoned when serious problems affected the station during and after the time it was launched into space atop a Saturn v carrier rocket.

Mounted where the third stage of a Saturn v would normally be located in the stack, the Skylab station lifted off at 5:30 p.m. on 14 May 1973 from Launch Complex 39A at Kennedy Space Center. Following separation, it slipped into a near-perfect orbit at 436 km (271 mi.) altitude. As scheduled, fifteen minutes after launch, the payload shroud surrounding the station was jettisoned.

As ground controllers checked for any anomalies, they found that most of the critical set-up tasks had been accomplished. In one, the Apollo Telescope Mount (a solar observatory) had rotated 90 degrees into its service position, and the four solar panels attached to it had also been deployed. While this major hurdle had been overcome, some irregularities were soon being detected and monitored. Of immediate concern was telemetry indicating that the protective heat shield, which also acted as a micrometeoroid shield, had not deployed. As a result, temperatures within the workshop were soaring. Additionally, there

The launch of the Skylab space station, 14 May 1973.

was a continuing lack of response to signals sent that should have activated the deployment of the main solar arrays, located on the exterior sides of the workshop section. Skylab's power production was therefore substantially lower than specification levels.

The cause of these major problems soon began to emerge, traced to severe acceleration forces during ascent from the launch pad. The micrometeoroid shield, whose function was to protect the crew against penetration by space debris and shade the workshop, had torn free. When this occurred, the shield also ripped off one of the two solar arrays mounted on either side of the workshop. As an outcome, the remaining array became entangled by debris that would prevent it from fully deploying once in orbit. While Skylab had successfully achieved orbit, and was theoretically ready for occupancy by the first crew, in its present condition the station was basically untenable, with interior temperatures hovering around 77°C (170°F).

The controllers were eventually able to orient Skylab so the smallest possible surface area was exposed to the Sun, which helped lower the interior temperature a little. But NASA now had to decide whether the troubled station should be abandoned or if they might contrive some way by which the first crew could be trained to salvage the situation through numerous untried procedures, which would likely include several potentially hazardous EVAS. One of these excursions outside Skylab would involve rigging up a makeshift micrometeorite shield to replace the one that had been torn off, and then freeing the partly deployed solar panel to allow full power to flow through the station. The risk factor was undoubtedly high, but the crew pledged their support and planning could continue.

Everyone connected to the Skylab programme – engineers, technicians, astronauts – now began looking at ways of creating and constructing workable, innovative ways of saving Skylab. With time pressing against them, this marathon, collective effort would later become known as NASA's 'Eleven Days in May'.

One major problem could potentially be solved using a small airlock that had been built into the workshop, designed to provide access to the vacuum of space for some planned experiments. A large collapsible Mylar parasol at the end of an 8-metre (26 ft) boom was developed which could be thrust out of this airlock and deployed over the side of the station that was mostly exposed to the Sun. This procedure was tested by the crew in a neutral-buoyancy simulator pool, and once refinements had been completed the parasol assembly was sent over to the Kennedy Space Center, ready for launch on the first crewed mission to Skylab.

The other problem, the badly tangled, jammed solar panel, fell to engineers at the Marshall Space Flight Center in Huntsville, Alabama, and they came up with a simple but practical solution. The answer was based on an everyday, pole-mounted tree pruning device like those sold

The first crew (SL-2) assigned to inhabit the Skylab station. From left: Joe Kerwin, Pete Conrad and Paul Weitz.

in hardware stores. The engineers set down a few modifications to this implement and contacted a local manufacturing company, who constructed a heavy-duty cable-cutting tool based on the tree pruner and mounted it on a collapsible 3-metre (10 ft) pole. This too was sent to Kennedy Space Center, and crew member Paul Weitz received special training for a stand-up cable-cutting EVA while positioned in the open hatch of the Apollo Command Module.[8]

### Skylab Back in Service

On 25 May 1973, ten days after originally planned, the Skylab 2 (SL-2) crew were launched within their Apollo spacecraft above a Saturn IB rocket. All went well, and the crew of Charles 'Pete' Conrad, Joseph Kerwin and Paul Weitz achieved orbit a little under ten minutes after lift-off. Conrad had soon guided their spacecraft onto a rendezvous trajectory. Seven hours and 40 minutes later they were closing in on the crippled Skylab station.

After conducting a fly-around of Skylab, Conrad confirmed to Houston that the micrometeoroid shield was missing, as was one of the solar panels, used to convert sunlight into electricity to power the station. Another solar panel was only partially deployed, tangled with debris from the lost shield, which included a thin aluminium strip wedged against the array cover. With all three astronauts fully protected in their space suits and sealed helmets, Conrad carefully manoeuvred the command module close to the station, at which time Weitz and Kerwin popped the spacecraft hatch. As Kerwin took a tight hold on his legs, Weitz stood up through the open hatch holding the heavy-duty cable-cutter. Over the next 40 minutes he tried unsuccessfully to cut away the debris, but the array remained stubbornly jammed. Exhaustion finally forced Weitz to cease his efforts. He withdrew back into the spacecraft and closed the hatch behind him.

Further frustration was in store for the tired astronauts when a docking was attempted and the Apollo docking probe failed to operate. Following the third attempt, mission controllers told Conrad they believed the fault lay in switches that controlled the catch activators. In order to fix this problem, the crew had to once again ensure they were fully suited up before depressurizing the spacecraft in order to connect a lead allowing them to bypass the switches, basically 'hot-wiring' the system. On the next attempt, while flying over the Pacific Ocean, the catches worked perfectly, and a hard dock was finally achieved. 'Hey!' Conrad told ground controllers. 'We got a hard dock out of it!' 'Way to go,' replied CapCom Richard Truly, his voice almost drowned out by the relieved cheers of everyone gathered in Mission Control. In a later transmission, as he looked back on the past strenuous 21 hours, Conrad said, 'We've had our problems and you've had your problems. We have eaten dinner and would like to hit the pad [get some sleep]. We'll press on first thing in the morning.' Mission Control concurred, and the astronauts were able to settle back in their couches for a well-earned sleep.[9]

The following day, 26 May, after a long sleep, the crew belatedly entered the Skylab station through the connecting tunnel. Each of them was wearing a gas mask amid fears there might be a lingering poisonous gas build-up in the damaged station, potentially released from overheated wall linings. Once it had been confirmed to be safe, the three astronauts knuckled down to repairing Skylab, which was running on limited power. It was oppressively hot inside the station, and every twenty minutes the crew had to retreat into an airlock for what they called a 'cooling break'.

One of their first tasks was to deploy the huge 7.3 by 6.7-metre (24 by 22 ft) canopy of the parasol by thrusting it upwards through the solar scientific airlock. Next they unfurled the paper-thin canopy on its sectional aluminium struts, which poked out like tent posts. Having completed this task, they returned to the command module, where they would spend the next three days waiting to see if the space station had cooled down enough to make it habitable. Mission Control later reported that the sunshield appeared to be working effectively, with telemetry indicating a significant drop in the interior temperature. An expected further fall to comfortable levels would allow the three astronauts to finally move out of their cramped spacecraft.[10]

Once the temperature reached a constant 27°C (81°F), the crew entered Skylab and began activating all the station's systems in order to prepare the research laboratory for the commencement of normal orbital operations. The crew now predicted they could safely remain in orbit studying the heavens, Earth, and the response of the human body to prolonged microgravity. Conrad, Kerwin and Weitz also finished moving into their two-storey, three-bedroom house in space and telecast a news conference from the dining room, showing through its round picture window a sweeping view of the Pacific coast passing 438 km (272 mi.) below. 'I think we've overcome our problems,' Conrad said during the eight-minute conference. 'I'm looking forward to a successful flight in Skylab of 28 days. Right now, we're in pretty good shape.'[11] However, there was one major task still remaining ahead of them.

In one of the most elaborate repair operations ever carried out in space – matched only seventeen years later by the repair of the Hubble Space Telescope – the Skylab astronauts succeeded in releasing the remaining main solar panel in a dramatic spacewalk on 7 June. Assisted by Joe Kerwin, Pete Conrad installed a handrail onto the jammed panel, following which he was able to pull himself over to where the twisted aluminium strap had prevented the solar array from deploying fully. Using a pair of giant bolt-cutters, Conrad first cut through the strip, which partially freed the panel. He then tied a rope to the panel and manually hauled it out to the full 90-degree angle. This then allowed the batteries to begin charging, and crucial additional power started flowing into the Skylab station. The entire operation was hailed as a complete success, and Skylab was finally and fully functional. The three-man crew of Skylab 2 had not only brought the space station back to life but, by their magnificent and courageous efforts, saved the entire Skylab programme.

Over the next weeks, Weitz and Kerwin were focused on carrying out as much science and research as possible. By the end of their 28-day mission they had completed 81 per cent of the planned solar observations using the Apollo Telescope Mount; 88 per cent of Earth observations

Kerwin (left) and Conrad performing the EVA that finally freed the jammed solar array.

using the Earth Resources Experiment Package (a group of six remote sensor systems, sensing visible, infrared and microwave radiation, furnishing data to numerous investigators on the ground in the earth sciences and technology assessment), and more than 90 per cent of the medical experiments.

On 18 June, the Skylab 2 astronauts broke the 24-day endurance record previously set by the crew of the ill-fated Soyuz-11 crew two years earlier. Two days later, Conrad and Weitz carried out a 96-minute spacewalk to retrieve film canisters from the ATM for return to Earth and replaced them with fresh ones. This being one of their final tasks, the three astronauts began preparing to shut down Skylab until the second crew (Skylab 3) arrived in July.[12]

On 22 June, the crew climbed into their Apollo spacecraft, suited up, and, following the undocking, photographed Skylab during a flyaround inspection before firing the service module's engine to bring them home. They splashed down some 1,280 km (795 mi.) southwest of San Diego and around 10 km (6 mi.) from the prime recovery ship, USS *Ticonderoga*. The spacecraft was hoisted onto the aircraft carrier and three crewmen climbed out through the hatch, a little unsteady but able to walk unassisted.

### Crew Change-over

On 28 July 1973, the second crew to occupy the Skylab space station lifted off from the Kennedy Space Center on what was predicted to be a record-breaking stay of 56 days aboard the orbiting laboratory. The Skylab 3 crew comprised Alan Bean (who had walked on the Moon on Apollo 12), Owen Garriott and Jack Lousma.

The docking operation went smoothly, although a hydrazine propellant leak developed in one of the service module's four clusters (also known as quads) of four thruster jets. This was cause for some concern, as the Apollo craft needed to be in perfect condition for the return to Earth. Bean reported a drop in pressure and said they had seen a stream of snow-like particles outside the spacecraft. Despite this, the crew had no problem in making their way aboard the unoccupied space station.[13]

Their long-duration mission, which NASA had already extended to 59 days, got off to a poor start when Lousma fell ill with space adaptation syndrome, followed soon after by Garriott and Bean. Despite their nausea, which soon passed, they began to adapt to living aboard the voluminous space station. Six days later another potentially hazardous leak of highly toxic hydrazine propellant was detected in their spacecraft, causing deep concern back in Mission Control. NASA began working around the clock putting in place contingency plans for a rescue mission should the Skylab 3 crew find themselves unable to return home in their spacecraft, which had been plagued by propellant leaks. These could reduce the craft's attitude control system to a minimum flyable condition. Another Apollo spacecraft was being hurriedly fitted with five couches, and a rescue team comprising astronauts Don Lind and Vance Brand had begun preparatory training.[14]

Skylab had a second docking port, which would allow the SL-3 crew to board the rescue craft, with all five astronauts returning to Earth together. Despite all of the preparations, these plans were eventually deemed unnecessary and were abandoned when Bean's crew determined that the problem was not as serious as originally believed. Their spacecraft, they said, could be safely manoeuvred using only two of the four sets of working thruster quads.

Following a six-day delay caused by their earlier illness, the first planned spacewalk could proceed. As mission commander Bean remained in the hatch area monitoring their activities, Lousma and

The second resident Skylab crew (SL-3): Owen Garriott (left), Jack Lousma and Alan Bean.

Jack Lousma during the EVA to erect a new sunshade over Skylab's exterior.

Garriott floated out at the end of lifelines and erected a new, sturdier V-shaped sunshade over the existing one placed there by the first crew, which was crimped in one corner. They locked it in place, changed the film in four solar telescope cameras, set up an experiment and checked out a suspected leak in the station's cooling system which was found not to exist. They re-entered Skylab after a mammoth 6.5-hour space-walk, satisfied that there was no coolant leak, and they were soon able to report that the temperature in the station had been further reduced, making their occupancy far more comfortable in the long term.[15]

The SL-3 crew would spend 59 days in space, completing 858 orbits of Earth before a Pacific splashdown southwest of San Diego at the end of their mission. During their occupancy of Skylab, Lousma had successfully completed two spacewalks outside the station. Although illness had affected all three astronauts early on, they managed to catch up and completed all the scientific tasks set for them, as well as a number not originally scheduled.

## A Space 'Mutiny'

The crew of the third and last Skylab mission, SL-4, was made up of the first all-novice astronaut crew since Gemini 8 back in March 1966, although both commander Jerry Carr and Bill Pogue were highly experienced military test pilots. For his part, solar physicist Ed Gibson had undertaken his year-long jet pilot training when selected by NASA as a scientist-astronaut in June 1965, earning his U.S. Air Force wings. They would set off on a record 85-day visit to the orbiting laboratory, during which time they would take an unprecedented look at the recently discovered Comet Kohoutek as it looped around the Sun in late December.

After two frustrating postponements of the launch, originally set for 10 November 1973, the final Skylab mission began with a lift-off from Launch Complex 39B six days later atop a Saturn 1B rocket. After a flawless launch and a five-orbit chase through space, their Apollo spacecraft caught up with the station, ready for the docking. To Carr's frustration, the first two attempts at docking ended in failure. He lined up a third attempt and this time successfully manoeuvred the spacecraft into the docking port. Much to everyone's relief, he reported a hard dock. They would enter the Skylab station the next day.

While the crew expected to be busy throughout their tenancy of the space station, they soon found themselves confronted by a heavily packed work schedule, which allowed little time for eating, sleeping or relaxing. It seemed that every hour of every day of their twelve-week mission had been planned in advance, and results had to be obtained on time. As their fatigue and frustrations grew, all three men became increasingly testy.

Early in the occupancy of Skylab, Bill Pogue's nausea had caused him to vomit one time. In a case of poor judgement, Carr had only reported Pogue's feelings of nausea to the ground and added that he had taken some motion sickness medicine and missed a scheduled meal. He and Gibson had discussed this between them and both feared that NASA's doctors would overreact if they knew of the vomiting, so no mention was made of the episode. Ground control first learned about Pogue's actual condition on tape recordings transmitted to them with routine information sent on non-public communications links. The crew was unaware that there was a recording of the vomiting incident in the transmission, prompting a furious chief astronaut Alan Shepard, once he found out, to chide Carr for covering up the full extent of Pogue's illness. 'We think you made a fairly serious error in judgement in not letting us know of your condition,' Shepard snapped at Carr over an open communications line. 'We are on the ground here to help you along and we hope you will let us know if you have any problems up there again as soon as they happen.' Carr was contrite and apologetic. 'Okay, I agree with you,' he responded. 'It was a dumb decision.'[16]

The crew continued their research and station duties, including a spacewalk by Pogue and Gibson to repair an antenna and replace the film in the solar observatory. Their spacewalk set a new EVA record of 6 hours and 34 minutes. The research work on board continued at a demanding and unrelenting pace, and there were increasingly exasperated calls from the crew requesting more leisure time. NASA was unresponsive and seemed quite determined to keep the crew to the programmed schedule. On top of this, however, they kept adding further tasks and experiments for the crew.

Three days after Christmas, on 28 December, the fatigued and overworked crew slipped up when they failed to maintain constant radio contact with the ground for a single orbit. It was a simple mistake; each crew member was meant to take a turn at listening and responding to voice traffic from the ground, but that day they had inadvertently left all three radios off. It was an error caused by a heavy workload, and once they had realized what had happened the crew resumed normal radio contact. It was during a later exchange of communications in which they were free to air their grievances that NASA finally relented and allowed the crew some additional leisure time for the remainder of

The third and final Skylab crew (SL-4): Jerry Carr, Ed Gibson and Bill Pogue.

their mission. Unfortunately that single orbit of radio silence led some over-enthusiastic reporters to state that the crew had deliberately switched off their radios for a full day as a protest, which in turn led to incorrect (and still perpetuated) reports that the crew had 'mutinied'.

The next day, the crew trained their sensitive cameras on Comet Kohoutek in an attempt to determine how it fared under the intense thermal and gravitational stresses of the Sun. The following day, 30 December, Pogue and Gibson would conduct another spacewalk to get a clearer view of the comet as it passed the Sun at 21 million km (13 million mi.) – its closest encounter. The crew later spoke by radio transmitter with the comet's discoverer, Czech-born astronomer Luboš Kohoutek, at the Hamburg Observatory in what was then West Germany.[17]

Creating a bizarre sort of record, the sl-4 crew moved into 1974 and back again to 1973 a total of sixteen times by crossing the International Date Line once every 93 minutes. By 8 February, the sl-4 astronauts had completed all of their assigned primary tasks, as well as several additional tests and experiments, and it was time to depart Skylab. After a record 83 days, 4 hours, 38 minutes and 12 seconds in orbit they undocked from Skylab, and after completing several more orbits re-entered the atmosphere, splashing down in the Pacific southwest of San Diego, where their spacecraft was soon retrieved and transported to the helicopter carrier uss *Okinawa*.

Apart from extensive post-flight medical tests and debriefing sessions, some discussions with the crew members would undoubtedly have taken place behind closed doors, but no disciplinary action seems to have ever been actioned on the crew for any indiscretions during their Skylab mission. There is no mention of any such discussions in Jerry Carr's 2008 biography, *Around the World in 84 Days*.[18] None of the three astronauts would fly into space again, but, given a complete lack of future missions – apart from the 1975 Apollo–Soyuz Test Project – this was not unusual in the following years, when NASA was transitioning to the space shuttle and many of the pioneering astronauts had left NASA for other pursuits.

NASA had originally intended for further crews to occupy Skylab once the space shuttle programme began around 1977. Tentative plans called for one of the first visiting shuttle crews to attach engines that would boost the derelict station into a higher orbit, allowing later missions to refurbish and work aboard Skylab. This concept was later

This final photograph of the Skylab space station in orbit was taken by the SL-4 crew as they prepared for their return journey after a marathon 83 days in orbit.

abandoned, partly due to increased solar activity, which slightly increased the density of the atmosphere, thereby increasing the drag on the station as it orbited Earth. This would have caused Skylab's orbit to decay and burn up during re-entry well before the shuttle became operational, so these plans were scrapped. The space station eventually re-entered the atmosphere, and in its final hours ground controllers managed to adjust Skylab's orientation in order to minimize the risk of fragments falling into populated areas. Although the station then burned up and disintegrated over the Indian Ocean as planned, some large segments still managed to rain down over remote and fortunately unpopulated areas of Western Australia on 11 July 1979.

Back in January 1975 there had already been a sad sidebar to the Skylab saga when another savage budget cut saw the cancellation of a second proposed Skylab station (Skylab-B). Visiting crews would have included astronauts who had earlier trained for a possible mission to rescue the SL-3 crew. Don Lind was one of those astronauts. 'Vance Brand and I and Bill Lenoir were going to obviously fly on the second Skylab,' Lind later reflected.

> We'd been backup for the last two of the Skylab 1 flights, and Vance and I were the rescue crew, and so it was perfectly obvious to everybody that we would fly on the second Skylab, when again they reduced the budget, cancelled the flight, cut the vehicle in half with a welding torch, and it now sits in the aerospace museum [National Air and Space Museum] in Washington. I think it's the most expensive museum display in the world, two and a half billion dollars. So I spent six and a half years training for two flights that never flew.[19]

# 9

# RECOVERING THE SOYUZ/SALYUT MISSIONS

On 27 September 1973, just two days after the crew of Skylab 3 had splashed down to conclude their marathon 59-day visit to the U.S. space station, two cosmonauts were launched into orbit around Earth, the first Soviet crew since the loss of three Soyuz -11 cosmonauts returning from the Salyut-1 space station two years earlier.

The flight commander of Soyuz-12 was Air Force Colonel Vasily Lazarev, with Oleg Makarov as flight engineer. A brutal lesson had been learned from flying three cosmonauts unprotected by space suits, and crews were now limited to just two members wearing the bulky suits, which precluded any third occupant. In a marked departure from the previous practice – and for the first time since the Voskhod-2 mission in March 1965 – cosmonauts would now wear these pressure suits for lift-off, docking and re-entry.

Original plans for the Soyuz-12 mission called for the two cosmonauts to dock with the Salyut-2 space station, which was launched on 3 April 1973. It was strongly rumoured across Moscow that a two-man crew had been in training for a docking mission to Salyut-2 and that a launch was imminent. Then everything went quiet. It was later reported that the station's initial orbit was higher and more elliptical than that of its predecessor, Salyut-1, which may have been due to a faulty poor performance by the Proton-K carrier rocket. A large number of fragments were detected in the orbital path, which resulted in some speculation that the rocket's D-1 upper stage may have exploded. According to a report on Soviet space programmes from 1976 to 1980, prepared for the U.S. Senate in October 1984:

The real trouble came on April 14 when Salyut was reported to have undergone a 'catastrophic malfunction' which ripped off the solar panels and boom-mounted rendezvous radar and radio transponder, leaving the vehicle tumbling in space without telemetry return. The craft may have separated into many pieces, some large enough to be tracked, but most were rather small and decayed quickly. Either an explosion or a misfiring thruster were blamed, although the most widely held theory was that the D-L upper stage had exploded with its debris damaging the space station.[1]

The remaining hulk of the station was finally de-orbited on 28 May and burned up during re-entry. In attempting to cover up the failure, Soviet officials stated that any operations planned for Salyut-2 had been successfully completed, and that all on-board systems and science equipment had functioned normally. It would be some considerable time before all the facts behind the loss of the station were finally revealed.

On 27 September 1973 the crew originally assigned to the link-up mission, Lazarev and Makarov, were launched on a two-day flight. They would spend two days in orbit aboard the newly redesigned Soyuz 7K-T spacecraft conducting integrated checking and testing of improved systems aboard the craft, as well as performing manual and automatic steering under various flight conditions and carrying out spectrographic photography of some parts of Earth's surface. The two-day flight of Soyuz-12 came to a safe ending with a landing on 29 September, after what was later described as a 'flawless mission' by Soviet space chiefs. The following two-man flight, Soyuz-13, also carried out a successful eight-day orbital mission.

Another station, identified as Salyut-3 – around a quarter the size of NASA's Skylab space station – would be launched and achieve orbit on 26 June 1974. The first docking with Salyut-3 was managed by the crew of Soyuz-14, under the command of veteran Vostok cosmonaut Pavel Popovich. He was launched on 3 July 1974 together with flight engineer Yuri Artyukhin, and the two men would spend sixteen days in orbit before undocking and returning to a safe landing.

The following mission, Soyuz-15, was nowhere near as successful. The crew of Gennady Sarafanov and Lev Demin (the first grandfather to fly into space) failed to dock with Salyut-3 due to a malfunction in the Igla docking system, and they were forced to return home after a flight lasting just two days.

Despite the failure of the previous mission, there were no plans to dock Soyuz-16 to the Salyut station following its launch on 2 December 1974. Instead the spacecraft was identical to the one that was to be used in the joint u.s.-Soviet Apollo–Soyuz Test Project planned for the following year, with the main aim of the flight being to rehearse and simulate procedures for this historic international meeting in space. After six days in orbit, commander Anatoli Filipchenko and flight engineer Nikolai Rukavishnikov re-entered to a safe touchdown on the snow-covered steppes of Kazakhstan.

Within five weeks of the end of the Soyuz-16 mission, another crewed spacecraft headed into orbit, this time targeted to a docking with the Salyut-4 space station, which had been launched on 26 December 1974. Its predecessor, Salyut-3, was still in orbit, but would be deliberately de-orbited on 24 January 1975. The Soyuz-17 crew of commander Alexei Gubarev and civilian flight engineer Georgi Grechko were launched on 11 January and completed a successful docking with Salyut-4. They would occupy the station and return at the end of their thirty-day mission – the longest crewed Soviet space flight up to that time.

The Soviet space effort suffered a near-catastrophe following the launch of Soyuz-18 on 14 April 1975, carrying Vasily Lazarev and Oleg Makarov, who had earlier teamed up on Soyuz-12. Their objective was to dock with Salyut-4, and sixty days aboard the orbiting station were scheduled, but those plans would go badly awry. Soon after lift-off, the Soyuz carrier rocket began veering off course and was automatically shut down. At the same time, the Soyuz spacecraft was explosively set free from the errant rocket and made a safe landing some 1,600 km (1,000 mi.) downrange in the rugged, sparsely populated Altai Mountains of western Siberia. Although neither cosmonaut suffered injuries during the aborted launch, the Soviets were more concerned by the fact that, had the booster rocket fired just a few seconds longer, the spacecraft could have come down in China, creating a diplomatic nightmare.[2] It was subsequently decided to relaunch the mission but with a different crew, and the aborted flight became officially designated as Soyuz-18A.

Six weeks after the near-calamitous flight of Soyuz-18A, mission commander Pyotr Klimuk and flight engineer Vitaly Sevastyanov lifted off from the Baikonur launch pad on 24 May for a rendezvous and docking with the Salyut-4 space station. Their occupancy of the station

Oleg Makarov and Vasili Lazarev were assigned a second mission together, but Soyuz-18 nearly ended in disaster when their booster rocket exploded over the launch pad.

would actually overlap the next Soviet space shot – the launch of Soyuz-19 on a mission to dock with an American Apollo spacecraft. Klimuk and Sevastyanov would complete the longest Soviet space mission to that time, lasting 63 days in total.

## A Handshake in Space

In May 1972, as part of previous discussions on a number of cooperative undertakings between the United States and the Soviet Union, President Richard Nixon and Soviet leader Leonid Brezhnev formalized plans for a joint space mission, in which they agreed at a Moscow summit to send American and Soviet spacemen on a historic joint Earth orbital flight in 1975. The agreement climaxed more than eighteen months of technical discussions held between the space agencies of both nations and was signed on the second day of the summit, which produced, among other agreements, a pact to limit strategic weapons.

The proposal hammered out called for an Apollo spacecraft carrying three astronauts to rendezvous and dock with a Soviet Soyuz craft that had earlier been launched into orbit with a crew of two cosmonauts. A specially developed docking collar and compatible airlocks would allow the pressurization difference between the two spacecraft to be adjusted following the docking. This would then allow them to remain locked together for two days, during which time there would be ceremonial meetings between the astronauts and cosmonauts, and a few minor experiments to be conducted. It was to be more than just a symbolic gesture of détente between the two superpowers; it would also demonstrate a practical means of carrying out an international space rescue between two totally different types of spacecraft, should it become necessary. A provisional date for the flight was given as July 1975.

Donald 'Deke' Slayton had originally been selected to fly the second orbital Mercury mission back in 1962, but was grounded in March of that year when medical tests revealed he was suffering from an ailment known as atrial fibrillation, an irregular heart rhythm. It had been a cruel blow, but he stayed on with NASA as the space agency's chief of flight crew operations at the Manned Spacecraft Center and subsequently played a key role in selecting future mission crews. Over the years, he never stopped dreaming of one day flying into space himself, jogging and working out in the astronauts' gymnasium to keep fit, as well as giving up his beloved cigars. In 1972 he underwent a further series of medical tests, which he passed, and his active astronaut status was restored. Slayton had already heard talk of the Apollo–Soyuz Test Project (ASTP), and he began taking Russian lessons even before it was announced, and formally handed over responsibility for the selection of the ASTP crew to Manned Spacecraft Center director Chris Kraft, all the while hoping he would be one of those given the nod. Soon after the completion of the final lunar landing, Apollo 17, Kraft announced that the ASTP crew would comprise Tom Stafford as commander, Vance Brand as command module pilot and Deke Slayton as docking module pilot. It was not the command Slayton had hoped for, but he was ecstatic; he was finally going to fly into space, sixteen years after being selected as a Mercury astronaut. Their four-man astronaut support team was composed of Karol Bobko, Bob Crippen, Dick Truly and Bob Overmyer. On 24 May 1973 the names of the Soviet prime and backup crew members were announced. The prime crew would consist of Alexei Leonov, the first person to walk in space on the Voskhod-2 mission in 1965, and Valery Kubasov, who had flown on Soyuz-6 in 1969.

Over the next two years, teams of scientists, astronauts and cosmonauts began shuttling between the United States and Russia, allowing for a crucial exchange of technical information and training, and the development of new equipment and procedures. It was not an easy task at times, with decades of bitter competition and mistrust

The Joint U.S.-USSR crew for the Apollo–Soyuz Test Project (ASTP) mission.
At top, mission commanders Tom Stafford and Alexei Leonov. At front are
Deke Slayton, Vance Brand and Valery Kubasov.

to overcome. Despite a distinct caution between the two Cold War
adversaries, a spirit of camaraderie began to evolve, particularly among
the key players in this international project – the three astronauts and
two cosmonauts. It had been realized early on that the entire programme would not work unless the five men developed a strong feeling
of mutual respect, and this actually happened quite rapidly between
the two crews, with home visits and private parties taking place.
Friendships evolved, especially as they each had to learn the other's
language in order to communicate while in orbit. Leonov laughingly
suggested he had to learn two languages: English, and Stafford's broad
Oklahomski!

Due to a lack of sophistication in its design, the Soyuz would play
a relatively passive role in the orbital link-up, although everyone on
the American side was warned to be diplomatic in not mentioning
such shortcomings.

The great space chase began on 15 July 1975 when the three American
astronauts lifted off from Cape Canaveral in their Apollo command
and service module (CSM), equipped on its nose with the Apollo docking module (ADM) – basically an airlock with docking facilities at each
end to allow for crew transfers. They were in pursuit of the two-man

Soyuz-19 spacecraft, which had been launched seven and a half hours earlier from the Baikonur launch complex. It was the first ever space launch broadcast live on television across the Soviet Union.

Two days later, after a great deal of orbital manoeuvring, the Apollo CSM was rapidly closing in on Soyuz-19. As a precaution, all five men were wearing their pressure suits in case something went dramatically wrong during the docking, but it all worked to perfection. At 11:09 a.m. Houston time on 17 July, Alexei Leonov announced, 'Contact!' Tom Stafford followed moments later with, 'We also have a capture. We have succeeded. Everything is excellent.' Obviously relieved, Leonov responded, 'Soyuz and Apollo are shaking hands now.'[3]

Three hours later, once all the docking security checks had been carried out, Leonov and Kubasov opened the Soyuz hatch that led into the docking module. Stafford opened the other end of the docking module hatch seven minutes later, following which he and Leonov briefly embraced in a symbolic gesture of cooperation between the two space superpowers. 'Glad to see you,' Stafford remarked in Russian. 'Very, very happy to see you,' Leonov responded in English. Stafford and Slayton then drifted into the Soyuz spacecraft, where they greeted Valery Kubasov, leaving Vance Brand temporarily alone in the Apollo CSM. Following formal greetings from Soviet leader Leonid Brezhnev and U.S. president Gerald Ford, the astronauts and cosmonauts exchanged

Deke Slayton and Alexei Leonov enjoying a fun moment aboard the Apollo spacecraft.

The launch of Soyuz-19 (above) and Apollo (often referred to as AS-18) on 15 July 1975.

gifts, held up flags of the other nation for the television cameras, and performed some other ceremonial tasks before enjoying a small meal hosted by Leonov and Kubasov while seated around a small green table.[4]

Over the next two days the astronauts and cosmonauts paid regular visits to each other's spacecraft, providing television images for the people back on Earth while carrying out medical and scientific experiments, sharing meals and conducting televised interviews.

Eventually it came time to undock the two spacecraft. Following a successful separation, a six-second firing of the ship's control rockets carried the Apollo CSM away from Soyuz-19 into a slightly higher, slower orbit, 233.7 to 226.9 km (139–41 mi.) above Earth, enabling the Apollo crew to see the Soyuz craft nearly a kilometre below as the distance between them increased. While the Apollo crew would remain in orbit a further three days, the Soyuz crew began preparations for a return to Earth. They would land gently and triumphantly amid a cloud of dust on a Central Asian prairie, and both cosmonauts emerged a little shaky but in otherwise good health.

Things did not go quite as well for the Apollo crew following their otherwise successful re-entry three days later. As they descended through 15,000 m (50,000 ft) altitude, two switches in the Earth Landing System (ELS) that were meant to be selected were mistakenly left as they were. Soon after, the interior of the command module began filling with what the crew described post-flight as a 'brownish-yellow gas'. This later proved to be highly corrosive nitrogen tetroxide, used as an oxidizer in their attitude-control thrusters. The crew began to choke on the fumes over the final minutes of their descent. Once they had splashed down the three men reached for oxygen masks, but Brand's was not a good fit and he passed out. As soon as possible, with Navy frogmen coming to their assistance and inflation bags righting the spacecraft, Stafford opened their hatch, letting in fresh sea air. Following an uncomfortable night aboard USS *New Orleans*, the carrier docked at Pearl Harbor and the three men were driven to the Tripler Army Medical Center in Honolulu, where they remained under observation for 48 hours. After numerous checks and close monitoring of their vital signs, all were eventually released.

Although a massive step forward in détente between the two space powers, which would later lead to other cooperative ventures in space, the last Apollo mission had almost ended in tragedy. It had been a close call for the three NASA astronauts and the American space programme as a whole.[5]

### Continuing the Soyuz Programme

Following the celebrated U.S.–Soviet link-up in space on the ASTP mission, Soyuz spacecraft continued to penetrate the skies. Soyuz-20 was an unmanned spacecraft launched on 17 November 1975 to a completely robotic docking with Salyut-4 as a long-duration test of improvements carried out on the spacecraft's systems. Once this had been completed after three months in orbit, a command signal was transmitted to the spacecraft which initiated the undocking procedure. Soyuz-20 re-entered on 16 February 1976 and landed by parachute in Kazakhstan.

While the unoccupied Salyut-4 remained in Earth orbit another two years, the third and last Almaz space station (under the designation of Salyut-5) was launched on 22 June 1976 and would play host to two crews – Soyuz-21 and Soyuz-23. Another mission, Soyuz-22, was an Earth observation mission that did not link up with the space station. Soyuz-24, launched on 7 February 1977, was the final mission to the Salyut-5 station, which was de-orbited in August, six months later.

The following month, on 29 September, Salyut-6 was launched into orbit, and ten days later the Soyuz-25 mission of commander Vladimir Kovalyonok and flight engineer Valery Ryumin headed for a rendez-vous and link-up with the orbiting laboratory. Barely 24 hours into their flight, Soviet ground control ordered the docking to be aborted when, following five attempts, the crew was unable to engage the docking latches of the space station. With dwindling fuel supplies and battery power also running low, they were ordered to return, landing northwest of Tselinograd, Kazakhstan, on 11 October after spending just over two days in orbit. The abortive mission was the third time in a little over three years that Soviet cosmonauts had been forced back to Earth because of a failure in the Soyuz navigation or docking systems.

Much to the relief of Soviet space chiefs, the crew of Soyuz-26 was able to achieve a full docking with Salyut-6 following their launch on 10 December 1977. Commander Yuri Romanenko and flight engineer Georgi Grechko would occupy the station, joined a month later by Vladimir Dzhanibekov and Oleg Makarov, who docked their Soyuz-27 spacecraft to a second docking port. The TASS news agency glowingly called the double docking 'a major accomplishment of Soviet science and technology, opening up broad vistas for the future utilization of outer space in the interests of science and the national economy.'[6]

It was at this time that the lengthy Interkosmos series of week-long 'guest cosmonaut' missions also began, in which a number of superficially trained cosmonauts from Soviet bloc nations accompanied a Soviet commander on missions to dock with and occupy the Salyut-6 space station.

### Soviet Flights and More Records

In June 1980, using the advanced Soyuz T spacecraft, the Soviet Union began a series of crewed missions to the Salyut-6 and later Salyut-7 space stations. On the third of these flights, one of the three crew members was a woman – Svetlana Savitskaya, who became the second female Soviet cosmonaut after Valentina Tereshkova some nineteen years earlier. Savitskaya was crewed with commander Leonid Popov and flight engineer Aleksandr Serebrov on the Soyuz T-7 mission.

On the second day of their flight, 20 August 1982, they successfully docked with the orbiting Salyut-6 station, joining the resident crew of the Salyut-5 mission, Anatoli Berezovoi and Valentin Lebedev, who had been on board since 14 May. Following a week's work aboard Salyut-6, Popov, Serebrov and Savitskaya boarded the Soyuz T-5 spacecraft for the journey home, leaving their Soyuz T-7 spacecraft linked to the space station for the use of the other crew, who would complete

Svetlana Savitskaya with her Soyuz T-7 crew, Leonid Popov (left) and Aleksandr Serebrov.

a record-setting 211 days in space before returning home.[7] Svetlana Savitskaya would fly into space a second time on 17 July 1984, together with Vladimir Dzhanibekov and Igor Volk on the twelve-day Soyuz T-12/Salyut-7 mission, during which she became the first woman to conduct a spacewalk outside the space station, on 25 July. Over 3 hours and 35 minutes, under the watchful eye of commander Dzhanibekov, Savitskaya conducted soldering and welding experiments during her history-making EVA.[8]

## A New Breed of Astronaut

Original plans called for the U.S. space shuttle programme to begin before 1978, but there were numerous developmental delays, especially with the vehicle's main engines and fragile thermal tiles, as well as countless other components. Given the lengthy delays and setbacks, the group of available astronauts had thinned considerably, and NASA recognized the need to recruit a whole new cadre of astronauts, whose training would be specific to the space shuttle.

On 8 July 1976, the space agency issued a call for space shuttle astronaut candidates, with applications accepted until 30 June the following year. In making the announcement, NASA said it was 'committed to an affirmative action program with a goal of having qualified minorities and women among the newly selected astronaut candidates.

Therefore, minority and women candidates are encouraged to apply.'⁹ In the past, the crewing of Mercury, Gemini and Apollo spacecraft had been assigned to men with previous test-piloting experience, or scientists willing to be trained as jet pilots, and NASA had received a great deal of criticism for not selecting women and minority astronauts. Now, for the very first time, the barriers that had once existed were gone, and they were eligible (and openly encouraged) to apply.

By the due date, NASA had received 8,079 applications; 1,544 coming from women. There were 1,261 pilot applications and 6,818 for the newly announced position of mission specialist. Eventually the applications were processed down to just 208. Beginning in August 1977, the first of those applicants started arriving at the Johnson Space Center to commence a week of personal interviews, along with physical and psychiatric examinations.

On 16 January 1978, NASA administrator Robert A. Frosch announced the selection of the 35 candidates who would begin their astronaut candidate (ASCAN) training later that year. Included in that number were six women, plus the first three African American candidates and the first Asian American trainee. Of the 35 named, fifteen were in the category of pilot, and the other twenty as mission candidates.

NASA's announcement of the names also stated that the newest cadre of trainees would report to the Johnson Space Center on 1 July, joining

NASA's first group of women astronauts. Top row (from left): Kathy Sullivan, Shannon Lucid, Anna Fisher and Judy Resnik. Kneeling at front: Sally Ride and Rhea Seddon.

The first African American astronauts selected by NASA in 1978. From left: Ron McNair, Guion (Guy) Bluford and Fred Gregory.

27 other still active astronauts for two years of training (later reduced to a single year), and to begin flying shuttle missions once it became fully operational.[10]

## Guests of the Interkosmos Programme

The previously mentioned Interkosmos series of high-profile space flights was a widely publicized Russian space programme that became a spectacular propaganda tool for the Soviet Union in the waning years of communism. Promoted as an international 'research cosmonaut' programme, it was also designed as a means of showcasing solidarity – however tenuous – with nine participating Eastern bloc countries, who supplied qualified pilots to be trained in Moscow for week-long 'guest' missions on orbiting Salyut (and later Mir) space stations.

It began back in April 1967, when the military leaders of socialist countries allied to the Warsaw Pact, namely Bulgaria, Cuba, Czecho-slovakia, the German Democratic Republic, Hungary, Mongolia, Poland and Romania, declared their interest in uniting their efforts in space exploration and research under the administration of the Soviet Union. The Interkosmos programme kicked off with the launch of research satellites on Soviet rockets. It worked so well that nine years later, on 16 July 1976, a further agreement was reached for the Soviet Union to train and fly a number of guest cosmonauts from member nations aboard its Soyuz and Salyut spacecraft. Each participating nation was to supply two qualified jet pilot candidates, and they began arriving for training at the Star City cosmonaut training centre located outside of Moscow in December 1976.

The first group, reporting for training in March 1978, were two candidates each from Czechoslovakia, Poland and East Germany, while a second group arrived from Bulgaria, Cuba, Hungary, Mongolia and Romania. Rounding out the programme, two candidates from Vietnam would join them in early 1979. Each of the candidates would be assigned to a senior cosmonaut for specific mission training, following which the better candidate of the two would be named as the prime crew member, along with their Soviet mission commander. As this was mostly a propaganda exercise, the Interkosmonauts would perform a number of simple tests or experiments during their flight, but they would have no systems-related duties. In fact, the Czech pilot Vladimír Remek was said to have been known post-flight as the 'red-handed' spaceman. When asked why, he laughingly replied, 'Oh, that's easy. On Salyut, whenever I reached for a switch or a dial or something, the Russians shouted, "Don't touch that!" and slapped my hands.' It's unlikely to be true, but it's a good story.

On 2 March 1978, amid much fanfare emanating from Moscow, Captain Remek became the first non-Russian, non-American person to fly into space when he accompanied Colonel Alexei Gubarev aboard Soyuz-28 on a seven-day mission to the Salyut-6 space station, then occupied by cosmonauts Yuri Romanenko and Georgi Grechko. This guest cosmonaut flight created great interest around the world. From a propaganda point of view the programme was considered a great success, and would continue.

The second Interkosmos flight, Soyuz-30, was launched on 27 June 1978 with Polish cosmonaut-researcher and Air Force officer Mirosław Hermaszewski flying to Salyut-6 for a seven-day tenure along with mission commander Pyotr Klimuk. This time, the resident Soviet crew occupying the station were Vladimir Kovalyonok and Aleksandr Ivanchenkov.

Over the next three years there would be seven more Interkosmos flights to the Salyut-6 space station, with participating guest cosmonaut-researchers Sigmund Jähn (East Germany), Georgi Ivanov (Bulgaria),

Czech Interkosmos participant Vladimir Remek (centre) with Alexei Leonov (left) and mission commander Alexei Gubarev.

Bertalan Farkas (Hungary), Pham Tuan (Vietnam), Arnaldo Tamayo Méndez (Cuba), Jügderdemidiin Gürragchaa (Mongolia) and Dumitru Prunariu (Romania) accompanying their respective Soviet commanders. The only real crisis arose when the Soyuz-33 flight failed to dock with the Salyut-6 station, and the crew of Nikolai Rukavishnikov and Bulgarian Georgi Ivanov were forced to return early. With the exception of that shortened mission, it had been planned that all the Interkosmos flights to the space station and back would last approximately seven days, with a few variables thrown in, just to demonstrate there was no favouritism shown to participating nations.

The Soyuz-40 mission would mark the end of the initial Interkosmos programme, but with Salyut-7 now replacing the older space station, it was decided to extend the programme to include a number of other countries, although they were not truly Interkosmos flights and were instead labelled as 'international missions'. On Soyuz-40, Romanian

## Interkosmos Missions, 1978–88

| Soyuz Mission | Guest Nation | Prime Crew | Backup Crew | Launch Date | Space Station |
|---|---|---|---|---|---|
| Soyuz-28 | Czecho-slovakia | Alexei Gubarev Vladimír Remek | Nikolai Rukavishnikov Oldřich Pelčák | 3 February 1978 | Salyut-6 |
| Soyuz-30 | Poland | Pyotr Klimuk Mirosław Hermaszewski | Valery Kubasov Zenon Jankowski | 27 June 1978 | Salyut-6 |
| Soyuz-31 | East Germany | Valery Bykovsky Sigmund Jähn | Viktor Gorbatko Eberhard Köllner | 26 August 1978 | Salyut-6 |
| Soyuz-33 | Bulgaria | Nikolai Rukavishnikov Georgi Ivanov | Yuri Romanenko Aleksandr Aleksandrov | 10 April 1979 | Salyut-6 |
| Soyuz-36 | Hungary | Valery Kubasov Bertalan Farkas | Vladimir Dzhanibekov Béla Magyari | 26 May 1980 | Salyut-6 |
| Soyuz-37 | Vietnam | Viktor Gorbatko Pham Tuan | Valery Bykovsky Bui Thanh Liem | 23 July 1980 | Salyut-6 |
| Soyuz-38 | Cuba | Yuri Romanenko Arnaldo Tamayo Méndez | Yevgeny Khrunov José Armando López Falcón | 18 September 1980 | Salyut-6 |
| Soyuz-39 | Mongolia | Vladimir Dzhanibekov Jügderdemidiin Gürragchaa | Vladimir Lyakhov Maidarzhavyn Ganzorig | 23 March 1981 | Salyut-6 |
| Soyuz-40 | Romania | Leonid Popov Dumitru Prunariu | Yuri Romanenko Dumitru Dediu | 14 May 1981 | Salyut-6 |
| Soyuz T-6 | France | Vladimir Dzhanibekov Aleksandr Ivanchenkov Jean-Loup Chrétien | Leonid Kizim Vladimir Solovyov Patrick Baudry | 24 June 1982 | Salyut-7 |
| Soyuz T-11 | India | Yuri Malyshev Gennady Strekalov Rakesh Sharma | Anatoli Berezovoi Georgi Grechko Ravish Malhotra | 2 April 1984 | Salyut-7 |
| Soyuz TM-3 | Syria | Aleksandr Viktorenko Aleksandr Aleksandrov* Muhammed Faris | Anatoli Solovyov Viktor Savinykh Munir Habib | 22 July 1987 | Mir |
| Soyuz TM-5 | Bulgaria | Anatoli Solovyov Viktor Savinykh Aleksandr Aleksandrov | Vladimir Lyakhov Aleksandr Serebrov Krasimir Stoyanov | 6 July 1988 | Mir |
| Soyuz TM-6 | Afghani-stan | Vladimir Lyakhov Valeri Polyakov Abdul Ahad Mohmand | Anatoli Berezovoi German Arzamazov Mohammad Dauran | 29 August 1988 | Mir |
| Soyuz TM-7 | France | Aleksandr Volkov Sergei Krikalev Jean-Loup Chrétien | Aleksandr Viktorenko Aleksandr Serebrov Michel Tognini | 26 November 1988 | Mir |

*Soviet cosmonaut: not to be confused with Bulgarian cosmonaut of the same name

Dumitru Prunariu became the ninth guest cosmonaut-researcher from the Soviet bloc countries on a standard Soyuz flight. The subsequent missions, beginning in 1982 with French spaceman Jean-Loup Chrétien, would launch with a number of guest cosmonauts from different countries that had been willing to pay to have one of their nationals conduct research in orbit.

Only two of these later Interkosmos missions would travel to Salyut-7, in 1982 and 1984, following which they would dock with and occupy the Mir space station, launched into orbit on 20 February 1986. After four such missions, this series of international flights came to an end with Soyuz TM-7, carrying Chrétien on his second visit to a Soviet space station. As the original flight of a Bulgarian cosmonaut had failed to dock with Salyut-6, that nation was allocated a second space mission, this time successful, to the Mir station. Instead of the original guest researcher Georgi Ivanov, however, his place on the flight was taken by Ivanov's Soyuz-33 backup, Aleksandr Aleksandrov.[11]

In the course of the decade-long programme, several space 'firsts' were achieved: first non-Soviet, non-American space traveller (Vladimír Remek), first international crew (Remek and Gubarev), first Asian cosmonaut (Pham Tuan), first Black cosmonaut (Arnaldo Tamayo Méndez) and first Indian citizen cosmonaut (Rakesh Sharma).

### The Shuttle–Mir programme

The Soviet space station known as Mir (which translates as 'peace' or 'world') was not only the largest and most complex spacecraft constructed to that time but the first to be assembled in orbit out of several individual modules. The first of these was the core module (DOS-7), which was launched from the Baikonur complex on 19 February 1986. Mir's first resident crew of Leonid Kizim and Vladimir Solovyov arrived in mid-March 1986, remaining on board until 5 May 1986. Over the ensuing four years another three modules would be attached to the core component: Kvant-1 (astrophysics module) in 1987, Kvant-2 (augmentation module) in 1989 and the Kristall technology module in 1990. Other modules would follow later in the programme.

Over subsequent years, a number of Soyuz crews would dock with Mir, and occupancy of the massive space laboratory would become almost routine.[12] These flights would include several in the Interkosmos programme, carrying guest cosmonauts. Other notable visitors to Mir during this time included Japanese journalist Toyohiro Akiyama (Soyuz TM-11, in 1990) and Helen Sharman from the UK, chosen to fly on Soyuz TM-12 in 1991 as part of Project Juno.

Growing up in Sheffield, Helen Sharman took an interest in chemistry, gaining her degree from Sheffield University in 1984 and later a PhD from Birkbeck College in London. In 1987 she became a food chemist for the Mars Wrigley Confectionery company in Slough. Two years later, she heard an intriguing 'wanted' advertisement on the radio, asking for people to apply to become Britain's first astronaut: 'No experience necessary'. The selected person would fly to the space station Mir on a mission known as Project Juno. This programme was a collaboration designed to boost relations between the Soviet Union and Great Britain as part of Soviet president Mikhail Gorbachev's policy of glasnost (openness). Initial funding came from a consortium of private British sponsors including British Aerospace, Memorex, Interflora and ITV. The competition was fierce, with nearly 13,000 applicants, but Sharman had a solid background in research science and food technology, was fit and had an aptitude for foreign languages. Eventually she was one of the two remaining candidates and together with Army Major Timothy Mace was sent to Moscow for training, following which she was the one selected to fly the mission. Sharman would launch aboard the Soyuz TM-12 spacecraft on 18 May 1991, spending eight days in space including a week aboard the Mir station before returning to Earth aboard Soyuz TM-11 on 26 May.

Out of the more than one hundred cosmonauts who occupied Mir in its fifteen-year history, one of the most remarkable stories belongs to Russian Sergei Krikalev, selected as a cosmonaut in 1985. He first flew to Mir in 1988 on the Soyuz TM-7 mission, remaining on the station for close to six months. In May 1991 he journeyed to Mir for a second time, this time as one of three crew members aboard the Soyuz TM-12 mission. Krikalev was due to return in October, five months later, but when one of the next two missions was cancelled he agreed to extend his stay.

Back on the ground, the once powerful Soviet Union was in its death throes, with Mikhail Gorbachev's political power rapidly dwindling as tanks ploughed through the streets. Gorbachev bowed to the inevitable and resigned, and on 26 December 1991, the Soviet Union disintegrated. In the face of this massive upheaval, no one seemed quite sure what to do about the cosmonauts still on board Mir, namely Sergei Krikalev and Aleksandr Volkov. The place they had launched from, the Baikonur

Sergei Krikalev became known as 'the Last Soviet Citizen'.

Cosmodrome, was now located in the newly independent republic of Kazakhstan, as was the planned landing site. Their mission was extended even further while decisions were being made. There was still a Soyuz capsule docked to Mir which Krikalev and his Ukrainian comrade Volkov could use for a hasty return home, but they knew if they decided to abandon Mir it could easily mean the end of the space station, and so they opted to stay and wait for any news.

Finally, Krikalev got word that a new crew was being launched to Mir in March and he would be returning home. The replacement crew arrived and Krikalev was finally able to return to Earth aboard the Soyuz TM-13 spacecraft, along with Volkov and German cosmonaut-researcher Klaus-Dietrich Flade. They landed on 25 March 1992 near the city of Arkalyk in what was now the Republic of Kazakhstan. Krikalev and Volkov still had the insignia of the Soviet Union on their uniforms and were carrying their Communist Party membership cards in the pockets of their suits. They would quickly become known in the Western press as 'the last citizens of the Soviet Union'. For Krikalev, he found that during his ten months in space, the world had changed completely. Even his birth town, Leningrad, had reverted to its former name of Saint Petersburg.[13]

## Americans on Mir

In 1993 a collaborative agreement was reached between officials from NASA and the Russian space programme to allow American space shuttles to rendezvous and dock with the Mir space station, with mutual benefits for both parties. This pact would also permit Russian cosmonauts to fly on shuttle missions and NASA astronauts to work alongside their Russian counterparts during a lengthy stay aboard the modular Mir laboratory. This cooperation was seen as an essential forerunner to the later joint construction and long-term occupancy of the International Space Station (ISS). In addition, it provided Russia with some badly needed funding to allow their cash-strapped space programme to continue.

The Shuttle–Mir programme began in 1994 and continued through to its scheduled completion in 1998. Within those four years, much was accomplished and many significant space events took place for the very first time, including the first flight of a cosmonaut (Sergei Krikalev) aboard a U.S. space shuttle when he joined the crew of STS-60, launched aboard the space shuttle *Discovery* on 3 February 1994. Krikalev would later fly a second shuttle mission in December 1998, when he and NASA astronaut Robert Cabana became the first humans to enter the International Space Station from the shuttle *Endeavour*, turning on the lights in the U.S. module *Unity*.

Since U.S. astronauts would be appointed to long-duration missions aboard Mir, NASA asked for volunteers. For some this was a popular career move, but for others the daunting prospect of spending six months training in Moscow on a completely different space programme, while at the same time undertaking crash courses in the Russian language, was not something that appealed to them – particularly for many military personnel who had once considered the Russians their enemies and still harboured a lingering distrust of their former Cold War enemy. However, a former U.S. Marine captain and scientist, Norman Thagard, was keen to take on the task, and when he was launched aboard the Soyuz TM-21 spacecraft on 14 March 1995 he became the first American to hitch a ride in a Russian spacecraft, seated alongside cosmonauts

Vladimir Dezhurov and Gennadi Strekalov. Their Mir-18 mission would last a total of 115 days, during which they conducted 28 experiments.

On 27 June 1995, space shuttle *Atlantis* lifted off from the Kennedy Space Center launch pad carrying a crew of seven, which included two Russian cosmonauts, on the STS-71 mission. This would be the fourteenth mission for *Atlantis* and the one-hundredth U.S. crewed space flight. The primary mission objectives were to rendezvous and dock with the Mir station; conduct joint U.S.–Russian life science investigations; carry out a logistical resupply of Mir; deliver the two Russian cosmonauts – Anatoli Solovyev and Nikolai Budarin; and retrieve Thagard, Dezhurov and Strekalov for the return to Earth. NASA administrator Daniel Goldin referred to the flight as the beginning of 'a new era of friendship and cooperation' between the United States and Russia.

Two days later, *Atlantis* made a cautious approach to Mir some 395 km (245 mi.) above Central Asia. Over the next two hours, shuttle

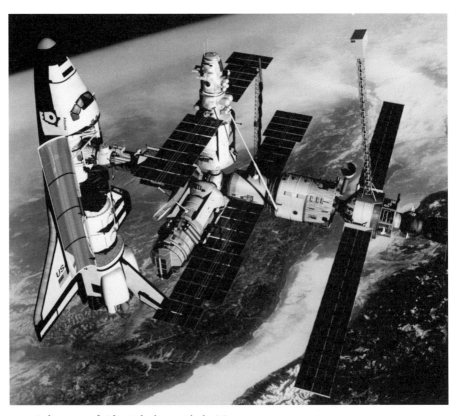

A depiction of *Atlantis* docking with the Mir space station on mission STS-71.

### NASA Astronauts in Residence on Mir Space Station

| Astronaut | Arrival at Mir | Departure Date | Days on Mir |
|---|---|---|---|
| Norman Thagard | 16 March 1995 | 29 June 1995 | 115 |
| Shannon Lucid | 24 March 1996 | 19 September 1996 | 179 |
| John Blaha | 19 September 1996 | 15 January 1997 | 118 |
| Jerry Linenger | 15 January 1997 | 17 May 1997 | 122 |
| Michael Foale | 17 May 1997 | 28 September 1997 | 134 |
| David Wolf | 28 September 1997 | 29 January 1998 | 119 |
| Andy Thomas | 25 January 1998 | 4 June 1998 | 130 |

commander Robert 'Hoot' Gibson slowly guided *Atlantis* towards the Mir station at a closing speed of 30 cm (12 in.) every 10 seconds. Just two seconds from the planned event timetable, *Atlantis* docked with the 112-tonne (123-ton) Mir station, thus creating the largest spacecraft in Earth orbit to that time.

*Atlantis* would remain docked with Mir for another five days before disengaging from the space station and returning to Earth with Thagard and his two Soyuz crewmates, leaving Solovyov and Budarin as the new resident crew aboard Mir. From February 1994 to June 1998, NASA would send eleven shuttle missions to Mir, with seven u.s. astronauts spending extended stays on the space station.

On the STS-76 mission, launched on 22 March 1996, Shannon Lucid was one of six NASA astronauts on board the third shuttle mission to dock with Mir, and she would remain on board the station when the other five crew members departed on 31 March and re-entered. Lucid eventually completed 179 days of science and related activities aboard the Russian space station before returning home aboard *Atlantis* on 26 September 1996 with the crew of the STS-79 mission.

### Problems with Mir

Between its launch in 1986 and its eventual demise in 2000, Space Station Mir played host to 104 visitors from numerous countries: 44 from the United States, 42 from the Soviet Union/Russia, six from France, four from Germany, and one each from Afghanistan, Austria, Bulgaria, Canada, Japan, Slovakia, Syria and the United Kingdom. Of that number, eleven of the people who boarded Mir were women.

While Mir was renowned for the tremendous work carried out in the station over its fifteen-year history, it also suffered from some close calls, which, at the least, could have meant sustaining damage through to placing some crew members in extreme danger. One of the most frightening hazards for any orbiting crew is fire. In February 1997, a fifteen-minute fire in an oxygen-generating device aboard Mir could easily have destroyed the space station and cost the lives of the then resident crew of six, which included NASA astronaut and physician Jerry Linenger.

On 23 February, during a routine ignition of what was later found to be a faulty lithium perchlorate canister – used to generate oxygen on Mir – a fire suddenly broke out inside the Kvant module. Alarms went off as flames shot outwards like a blowtorch and molten metal began exuding from the burning canister, filling the station with thick, choking smoke. 'I did not expect smoke to spread so quickly,' Linenger later recalled. '[It] was about 10 times faster than I would expect a fire to spread on a space station.'[14]

The dense smoke also blocked one of the two routes to the two docked Soyuz spacecraft that could have been used as an emergency 'lifeboat' in an evacuation of the station. This blockage effectively meant that only three of the six crew members on board could have left the stricken spacecraft. Some of the crew managed to don respirator masks, although Linenger's did not work and he had to don another one to fight the fire. They then took turns using three extinguishers to douse the flames. This proved quite difficult, as the lack of gravity meant that the firefighter had to be held in place by another crew member to prevent them being propelled backwards with the Newtonian reaction. Linenger later estimated that the fire burned for about fourteen minutes before it was finally extinguished.

Once the smoke had dissipated somewhat, Linenger examined his five crewmates. Fortunately, none had suffered any serious effects from smoke inhalation. Remarkably, too, there was no irreparable damage to the station, and it subsequently continued to function as normal. Following a crew changeover, Jerry Linenger returned to Earth with the STS-84 crew of *Atlantis* on 24 May 1997, after a total 132 days in orbit.

Just weeks later there was another serious incident. In April 1997 an unmanned cargo ship, Progress M-34, had been launched to a successful rendezvous and docking at the aft port of Mir's Kvant-1 module. Two months later, on 24 June, a planned docking test began with the Progress craft detaching from the module under manual control and holding station nearby. The following day, crew members Aleksandr Lazutkin and NASA astronaut Michael Foale watched with interest as Vasily Tsibliev fired the cargo ship's rocket and carefully guided Progress back towards the docking port using a remote system, while monitoring the manoeuvre on a video screen, which showed an image relayed from a camera on board Progress. Interest quickly turned to panic as Tsibliev unexpectedly lost control of the rapidly approaching vehicle. He fired the braking rockets, but far too late.

Lazutkin later described the onrushing Progress as looking 'full of menace, like a shark'. He added, 'I watched this black body covered in spots sliding past below me. I looked closer, and at that point there was a great thump and the whole station shook.'[15] The 7.15-tonne (15,800 lb) spacecraft had slammed into a solar array mounted on the Spektr (Spectrum) module before hitting the Spektr module itself, punching a hole in a solar panel, buckling a radiator and breaching the integrity of Spektr's hull. Moments later, the crew could hear the ominous hissing sound of air escaping from the station and felt their ears pop. The collision had caused a power outage, so in the darkness they had to scramble to find cutting tools, with which they severed lines leading into the stricken Spektr module, and safely sealed the hatches. Meanwhile, Tsibliev manoeuvred the Progress craft – later de-orbited – away from the now spinning station. Ground control managed to fire rockets to control the spin and point the undamaged solar arrays towards the Sun, restoring light and power.

Despite these near disasters and ongoing computer and power outage problems and air leaks, crews continued to operate aboard Mir. However, the problems of 1997 were not yet over. In September, the three resident crew members, now including British-born NASA astronaut Michael Foale, had to seal themselves inside their Soyuz spacecraft ready to initiate a quick emergency undocking as an American military satellite bore down on the space station. It was projected to miss Mir, but only by a small margin (later estimated to be less than a kilometre), and once the all-clear was given the three crew members were able to leave their Soyuz craft and return to life aboard Mir.

All too soon, NASA began exerting pressure on a cash-strapped Russia to meet its financial commitments to the new International Space Station, rather than reinvesting in the problem-plagued and ageing Mir, which had suffered a chequered history, including 1,600 equipment

In all, seven U.S. astronauts would spend long-duration stays aboard the Mir station. They were (front, from left) Norman Thagard, John Blaha, Jerry Linenger and David Wolf. At rear: Andy Thomas, Shannon Lucid and Michael Foale.

failures as well as the critical docking collision and terrifying on-board fire. On 23 March 2001, the end came for Mir when Russia's space agency allowed the station to de-orbit, which resulted in the 122-tonne (134-ton) structure breaking up in the ferocious heat of re-entry. The remaining debris crashed into the South Pacific.

# 10

# SPACE SHUTTLES AND THE ISS

It was one of the most awe-inspiring yet fraught creations ever devised and constructed. The space shuttle was a magnificent winged spacecraft that could fly into orbit before returning to an unpowered landing on a conventional runway. From 1981, a fleet of five shuttles – or orbiters as they were also designated by NASA – completed more than 130 missions. It became the workhorse of America's space programme for three decades, a superhuman feat of engineering that captivated millions around the world.

But for all the elation associated with its many triumphs, the failures of the space shuttle were deeply heartbreaking, most notably when the *Challenger* orbiter and its crew of seven were lost 73 seconds into its tenth launch, on 28 January 1986. Then, just seventeen years later, on 1 February 2003, yet another crew of seven astronauts perished when NASA's first operational shuttle, *Columbia*, disintegrated during its atmospheric re-entry.

Exploration often comes at a cruel cost, and NASA was not exempt from the disasters that can happen when human beings defy the immutable laws of gravity and venture beyond the atmosphere on journeys into the unforgiving perils of space. Nevertheless, America's space shuttle retains an amazing history of human endeavour and ingenuity.

### A Remarkable Machine

'Hello Houston, *Columbia* here!' Veteran astronaut and mission commander John Young sounded amazingly calm when his transmission finally broke through after a fifteen-minute communications blackout. The men and women of Mission Control in Houston erupted into spontaneous cries of joy, pride and relief. They shook hands, clapped each other on the back and broke out the celebratory cigars. Minutes later, they watched on screens as the heat-scarred space shuttle swept out of the blue skies and glided towards the dry lakebed runway at Edwards Air Force Base in the Mojave Desert.

As *Columbia* decelerated, a dramatic double sonic boom thundered across the desert. 'What a way to come to California!' laughed pilot Bob Crippen. Just five minutes later, the orbiter's main wheels touched down and the half-million spectators gathered at the landing site went wild. Nine seconds later, the ship's nose wheels also settled on the runway, and soon after America's newest space vehicle had rolled to a stop in the shimmering desert air. 'Welcome home, *Columbia*,' came an exuberant voice from Mission Control. 'Beautiful, beautiful!'

The first space shuttle's return marked a fitting end to the pioneering phase of America's space programme. Gone were the great, candy-striped parachutes and single-use spacecraft splashing down into the sea, followed by a helicopter recovery. Instead, a whole new age of space travel had dawned – an era of reusable spacecraft designed to fly dozens of missions and land like an aircraft on a regular runway.

Incredible progress had been made in the first two decades of human space flight, but the United States was aiming higher. *Columbia* was far more than just a winged vehicle of astonishing complexity. She provided a much-needed affirmation of America's technological prowess at a time when the nation's citizens were questioning that very capability. There had been disheartening defeats in the unpopular Vietnam War, which had ended only six years earlier. More recently, American rescue helicopters had been downed in the Iranian desert; there had been the debacle of the partial core meltdown at the Three Mile Island nuclear plant; and automobile capital Detroit was still reeling from the onslaught of imported Asian cars.

Although a prototype shuttle – named *Enterprise*, after Captain Kirk's fictional *Star Trek* spacecraft – had taken part in unpowered approach and landing tests from atop a specially modified Boeing 747 carrier aircraft, *Columbia*'s maiden flight would be the first time a shuttle had ever been sent into space. The flight would be known as STS-1, the maiden mission of a Space Transportation System vehicle, as the space shuttle was officially designated, and two of NASA's finest

The crew of the first Space Shuttle orbital mission, STS-1, John Young (left) and Bob Crippen.

astronauts were at the helm. Fifty-year-old commander John Young was making his fifth space flight, having flown two successful Gemini missions and then journeyed twice to the Moon during the Apollo programme. His 43-year-old pilot was rookie Bob Crippen, a Navy aviator who had been transferred to NASA's astronaut corps from the U.S. Air Force's defunct Manned Orbiting Laboratory (MOL) programme.

The principal mission objectives were to demonstrate the safe launch and return of the shuttle and her crew, and to verify the combined performance of the entire shuttle 'stack', as it was known – the space shuttle, solid rocket boosters and external fuel tank. There were only two payloads on board: the Development Flight Instrumentation (DFI) and the Aerodynamic Coefficient Identifications Package (ACIP) pallet, each containing sensors and measuring devices for recording temperatures, stresses and acceleration levels.

STS-1 would be the first and so far only maiden test flight of a new orbital spacecraft to lift off with a crew on board. So for this and the following three test flights, *Columbia* was fitted with the kind of ejection seats that had been developed for the supersonic Lockheed SR-71 'Blackbird' sky plane. They would provide Young and Crippen with some chance of evacuation in the event of a crisis during ascent (up to a speed of around Mach 3) or if the orbiter was nearing subsonic speeds following re-entry but could not reach a runway. After the test flights were completed and the shuttle was declared operational, the ejection seats were considered too heavy to warrant providing one to each member of the larger crews, so the two already fitted for the pilots were deactivated. After the ninth mission they were removed entirely. The astronauts knew that any emergency ejection from a shuttle at anything in excess of subsonic speed during ascent or descent would prove fatal. As well, the mid-deck seats could not be fitted with ejection devices. In the later loss of shuttles *Challenger* and *Columbia*, the crew would have had only milliseconds to react, so ejection seats would not have saved any crew members, even if fitted. Furthermore, in a hypothetical situation, any crew member ejecting prior to the discarding of the two solid rocket boosters would have ended up in the midst of a white-hot fire trail. Even after booster deployment, the shuttle would have been far too high and accelerating at too massive a rate to eject. When *Columbia* disintegrated it was tearing through Earth's mesosphere at a non-survivable 61 km (200,000 ft) altitude and barrelling along at eighteen times the speed of sound. In both instances, any crew member ejecting would have been instantly torn apart in the colossal wind blast associated with supersonic ascent and descent.

The space shuttle itself was about the size of an Airbus A320 airliner, but with double-delta wings. At just over 37 m (122 ft) long, it was the same length as the first powered flight made by Orville Wright in December 1903. The shuttle's main body was mostly comprised of aluminium alloy, while the engine structure was primarily made up of titanium alloy. The underbelly of the orbiter's fuselage was covered with individually crafted ceramic tiles designed to protect the vehicle during the ferocious heat of re-entry.

Each orbiter had three main engines, fed by a massive external fuel tank. At maximum power they delivered over 27,500 megawatts of power, or 37 million horsepower. The liquid hydrogen that powered them was maintained at a temperature of $-252°C$ ($-422°F$), yet the temperature in the engines' combustion chamber reached $3,350°C$ ($6,000°F$), or 500 degrees higher than the boiling point of iron. The high-pressure turbopumps feeding the engines spun at 37,000 rpm. One of them could fire a column of liquid nitrogen 58 km (36 mi.) into the air.

*Orbiter Specifications*
Length: 37.237 m (122 ft)
Wingspan: 23.79 m (78 ft)
Height: 17.86 m (58 ½ ft)
Empty weight: 78,000 kg (172,000 lb)
Gross lift-off weight: 110,000 kg (242,000 lb)
Payload bay dimensions: 4.6 × 18 m (15 × 59 ft)

## Lift-off!

Purely by chance, *Columbia*'s day of fame took place on 12 April 1981, twenty years to the day after Yuri Gagarin's epic flight into the cosmos. Two days earlier, the countdown had been bedevilled by a number of technical issues and a timing skew in the orbiter's general-purpose computer system, which had combined to delay the lift-off. The launch would eventually take place from Launch Pad 39A at the Kennedy Space Center, at three seconds after 7:00 a.m. EST.

It came as no surprise to engineers at the Cape that *Columbia*'s lift-off made a terrible mess of the launch pad. White-hot flames gushing from beneath the orbiter's solid rocket boosters created temperatures of 3,300°C (5,970°F), charring cables, melting water-spray nozzles and searing grass more than a kilometre from the pad. The massive shock wave damaged light fixtures and alarm boxes. Electrical panel doors were torn from their fittings, and elevator doors were buckled out of their frames. Nevertheless, a post-launch inspection showed that the overall damage was actually well within expectations.

Barely 45 minutes after lift-off, *Columbia* was circling the planet at an altitude of 240 km (149 mi.), which would increase to 270 km (168 mi.) by the end of the day. There were only a few minor problems. During their first full 'night' in space the two astronauts complained about a chill in the cabin, which was remedied when a signal from ground controllers pumped warm water into the cabin's temperature control unit. Young and Crippen were also unable to repair a faulty flight data recorder, as the cover had been screwed down too tightly on the ground – they could not muster sufficient torque in weightless conditions to undo the screws.

The most serious problem occurred on the second night when an alarm bell rang out and a warning light began flashing. The alarms signalled a malfunction in a heating unit on one of the auxiliary power

Launch of *Columbia* on the first shuttle orbital mission, 12 April 1981.

units for *Columbia's* hydraulic systems, which controlled the landing gear and the elevons (trailing edge flaps) in the wings. The problem was quickly fixed with a simple flick of a switch. The two astronauts also noticed that a number of protective heat tiles were missing from the rear of the orbiter, clearly dislodged during the launch. After consultation with the ground it was decided that the lost tiles were from non-critical areas, and their absence should not affect a safe re-entry.

Although no serious scientific work was carried out during the flight, most of *Columbia's* systems were checked, with a double opening and closing of the payload bay doors confirming the integrity of their operation in weightless conditions. On the second day, a little over an hour before *Columbia* was scheduled to land, Young and Crippen punched a series of commands into the computer's guidance system, which would control most of the re-entry phase. Small thruster rockets came to life in a series of short 'burns' that rotated the spacecraft through 180 degrees, before the main engines fired for 147 seconds in a prolonged deceleration burn. Twenty minutes later, and now travelling at 27,200 km/h (17,000 mph), *Columbia* began the most crucial and agonizing minutes of her descent. With radio communications blacked out by electromagnetic interference, the craft plunged into the searing 1,650°C (3,000°F) heat of atmospheric re-entry.

Over the next seventeen minutes, flight controllers waited for the interference to subside before John Young's calm salutation broke both the radio silence and the tension. *Columbia* then changed roles, from spacecraft to unpowered glider. Her computers swirled the orbiter through a deliberate series of turns designed to wash off some of her speed, and Mission Control told the two astronauts they were 'right on the money'. At about 3,700 m (12,000 ft), Young took over the landing, operating the flaps, elevons, rudders and speed brakes for the tricky final approach. He knew he only had one chance at getting *Columbia* down, which he accomplished with great precision, gliding to a touch-down at around 345 km/h (214 mph) – nearly 50 km/h (31 mph) faster than a commercial jetliner.

A post-flight inspection revealed that a massive pressure wave had been created during the launch, following the ignition of the solid rocket

### Space Shuttle *Columbia* Orbital Test Flights

| Mission | Prime Crew | Backup Crew | Launch Date | Landing Date | Orbits |
|---------|-----------|-------------|-------------|--------------|--------|
| STS-1 | John W. Young<br>Robert L. Crippen | Joe H. Engle<br>Richard Truly | 12 April 1981 | 14 April 1981 | 36 |
| STS-2 | Joe H. Engle<br>Richard H. Truly | Thomas K. Mattingly II<br>Henry W. Hartsfield Jr | 12 November 1981 | 14 November 1981 | 37 |
| STS-3* | Jack R. Lousma<br>C. Gordon Fullerton | Thomas K. Mattingly II<br>Henry W. Hartsfield Jr | 22 March 1982 | 30 March 1982 | 130 |
| STS-4 | Thomas K. Mattingly II<br>Henry W. Hartsfield Jr | — | 27 June 1982 | 4 July 1982 | 113 |

*All test flights landed on Runway 23 at Edwards Air Force Base, except for STS-3, which landed on Runway 17 at White Sands Missile Range, New Mexico, due to flooding at Edwards.

boosters. It was this wave that had shaken loose sixteen tiles and damaged another 148. In all other respects, though, the orbiter had completed her maiden flight with flying colours.

The success of *Columbia*'s first flight lifted Americans out of a collective sense of futility and gloom, finally giving them something to cheer about – an innovative space vehicle flown by two incredibly brave astronauts. As a relieved President Ronald Reagan told Young and Crippen: 'Through you, we feel as giants again.'[1]

Because *Columbia* was heavier than the other orbiters, it became something of a workhorse of the shuttle fleet, hauling experiments into space on extended science missions. Among her many achievements, however, were the deployment of the Chandra X-Ray Observatory and the servicing of the Hubble Space Telescope.

## Shuttle Flights Continue

With the successful completion of the orbital test missions, NASA declared the space shuttle fully operational, and the next two flights, STS-5 and STS-6, carried crews of four. On the first of these flights, crewed by mission commander Vance Brand, pilot Robert Overmyer and mission specialists Joe Allen and Bill Lenoir, NASA was able to demonstrate the business potential of the orbiter by transporting two communications satellites into space in the voluminous payload bay of orbiter *Columbia* on its fourth journey into orbit. The satellites were successfully deployed by the crew, and the only real mission failure came when a planned spacewalk by Allen and Lenoir had to be called off after a cooling fan in Allen's suit failed and could not be repaired. Lenoir suggested he could attempt a solo EVA, but this was vetoed by NASA. The five-day mission ended on 16 November 1982 with a perfect touchdown on the desert landing strip at Edwards Air Force Base, California.

On the following flight, STS-6, *Challenger* made her maiden journey into space following a flawless launch on 4 April 1983, carrying a crew of four: commander Paul Weitz, pilot Karol Bobko, and mission specialists Story Musgrave and Don Peterson. Donning new-style EVA space suits that had been specifically designed for the purpose, Musgrave and Peterson were able to accomplish the shuttle programme's first tethered spacewalk activity, spending 4 hours and 17 minutes conducting a number of tests in *Challenger*'s payload bay.

*Challenger* would also carry the next crew into orbit on the STS-7 mission, launched on 18 June 1983 on a six-day mission. The date was significant, in that exactly twenty years earlier the first woman to fly into space, Valentina Tereshkova, was midway through her historic orbital flight, and on this seventh shuttle mission the crew of five included America's first female astronaut, mission specialist Sally Ride. A 31-year-old physicist from California who had been a tennis champion in her teens, Ride and five other women had made space flight history five years earlier when they were accepted into NASA's space programme to train alongside their male counterparts for future shuttle missions. The crew's principal tasks involved the deployment of Canada's Anik satellite from the payload bay and, later, on the eighteenth orbit, the Palapa-B, an Indonesian communications satellite.

Physician and astronaut Norman Thagard had been a late inclusion in the STS-7 crew; he was on board to conduct medical tests surrounding the mystery malady known as space adaptation syndrome, which had caused extreme and unexpected nausea for some previous crew members – but curiously not others – early on in their missions. The cause of this lingering illness needed to be investigated by qualified practitioners in space and on the ground and, if possible, some sort of treatment found.

Mission Specialist Sally Ride became the first American woman in space aboard shuttle *Challenger* on the STS-7 mission.

On the fifth day of the flight, Sally Ride and John Fabian on STS-7 released the West German-built Shuttle Pallet Satellite (SPAS) using the shuttle's remote manipulator arm, and then practised both retrieving and releasing the experimental satellite. The STS-7 flight plan called for the first shuttle landing at the new 3-mile-long concrete runway at the Kennedy Space Center, but even as they prepared for re-entry a forecast of bad weather at the Cape caused them to fly an additional two orbits while waiting for a decision. Eventually the crew had to be denied a Florida landing, touching down instead to a lonely reception at Edwards Air Force Base. 'The thing that I'll remember most about the flight is that it was fun,' Ride later recalled. 'In fact I'm sure it was the most fun I'll ever have in my life.'[2]

The 35 shuttle astronauts selected in 1978 were certainly well represented in the early orbital missions; on STS-7, four of the five crew members were from the Group 8 cadre, and on the following flight, STS-8, three of the five also came from that group. Once again, there was a social-history first for NASA, when Guion 'Guy' Bluford became the first African American astronaut to fly into space.

To a lesser extent, another piece of shuttle history was recorded on 30 August 1983. It was not only the first night-time launch of a space shuttle but the first nocturnal launch since the midnight lift-off Apollo 17 back in 1972. The launch time for STS-8 was essential in accomplishing one of the main goals of the six-day mission: the deployment of an Indian communications and weather satellite, with orbital mechanics dictating when shuttle *Challenger* could be launched in order to be in the correct position to deploy the Insat-B satellite and send it on its way to an altitude of 35,900 km (22,300 mi.). The launch was successful, and on the third mission day, mission specialists Bluford and Dale Gardner tested the strength of the Canadian-built remote arm by deploying and then retrieving a 3,470-kilogram (7,640 lb) dumbbell-shaped test object made of aluminium, steel and lead that had been stowed in *Challenger*'s payload bay. Further experiments were also carried out by Dr William Thornton, following on from the work conducted by Thagard on the previous mission concerning the effects of the so-called 'space sickness'.[3] After a successful six-day flight, the STS-8 mission ended with a night-time landing – again a first for NASA – on the Edwards runway.

While the introduction of the space shuttle had brought forth a new crew designation of mission specialist, a further category of

The large crew of STS-9/Spacelab 1. Front (from left) MS Owen Garriott, pilot Brewster Shaw, commander John Young and MS Bob Parker. At back: PS Byron Lichtenberg and PS Ulf Merbold.

space traveller would be introduced on mission STS-9: that of payload specialist. Unlike NASA's mission specialists, these were men and (eventually) women who were selected by their own nations or organizations – even the military – and charged with hosting experiments, hardware and satellites that had to be delivered into space aboard a shuttle. They did not have to go through the same rigorous training process developed for NASA astronauts; instead they received an abbreviated training programme in Houston on basic shuttle systems, operations and safety procedures.

When *Columbia* thundered off its Florida launch pad on 28 November 1983 on the nine-day STS-9 mission (later extended to ten days), it carried the first two payload specialists. These were West German physicist Ulf Merbold from the European Space Agency (ESA), who became the first European to fly on one of NASA's shuttles, and American biomedical engineer Byron Lichtenberg from the Massachusetts Institute of Technology. They were on board to conduct experiments aboard the billion-dollar Spacelab, ESA's scientific research laboratory, secured like a large caravan inside *Columbia*'s voluminous payload bay, which had been built in West Germany under the auspices of ESA. Access to (or passage from) the Spacelab modular laboratory in the shuttle's

## Shuttle Flights under STS Numerical Designation

| Flight | Orbiter | Crew | Launch | Landing | Orbits |
|---|---|---|---|---|---|
| STS-5 | *Columbia* (OV-102) | Vance Brand Robert Overmyer Joseph Allen William Lenoir | 11 November 1982 | 16 November 1982 | 81 |
| STS-6 | *Challenger* (OV-099) | Paul Weitz Karol Bobko Story Musgrave Donald Peterson | 4 April 1983 | 9 April 1983 | 81 |
| STS-7 | *Challenger* (OV-099) | Robert Crippen Frederick Hauck John Fabian Sally Ride Norman Thagard | 18 June 1983 | 24 November 1983 | 97 |
| STS-8 | *Challenger* (OV-099) | Richard Truly Daniel Brandenstein Guion Bluford Dale Gardner William Thornton | 30 August 1983 | 5 September 1983 | 98 |
| STS-9/ Spacelab-1 (STS-41A) | *Columbia* (OV-102) | John Young Brewster Shaw Owen Garriott Robert Parker Ulf Merbold Byron Lichtenberg | 28 November 1983 | 8 December 1983 | 167 |

open payload bay was achieved by moving through a pressurized transfer tunnel located between the laboratory and a sealable hatch in the shuttle's mid-deck crew cabin. Once they had entered the Spacelab module, Merbold and Lichtenberg began conducting over seventy experiments designed by scientists from the United States, Western Europe, Japan and Canada. It promised to be the longest and most scientifically productive shuttle flight to that time, which proved to be the case.

While the flight was basically trouble-free, the latter part of the mission was beset by serious problems. On 8 December, commander John Young fired RCS thrusters to establish the correct shuttle orientation four hours before re-entry, at which point one of the flight control computers crashed, followed by a second a few minutes later. After rebooting the system, Young delayed the re-entry, allowing *Columbia* to drift in orbit for several hours while the crew double-checked the spacecraft's systems. Once the re-entry began it all went smoothly, ending with a touchdown on Runway 17 at Edwards. What the crew did not know until later was that around two minutes before landing, two of *Columbia*'s three auxiliary power units (APUs) had caught fire as the result of a hydrazine leak, which sprayed the toxic fuel onto a hot surface and ignited the two APUs. Fortunately for the crew, the fire burned itself out, while still causing major damage to the compartment; but NASA had just come within a whisker of losing a shuttle and its six-man crew in a landing explosion.[4]

### Adopting New Flight Designations

From the outset of the space shuttle programme, each mission was given a progressive numeric designation. The first orbital flight was STS-1, and this would continue until STS-9, when a new, more complex system was initiated. Under this system, STS-9 also had a second designation of STS-41A. The first number (in this case 4) would indicate the NASA fiscal year in which the mission was scheduled to fly, with each NASA fiscal year beginning in October. The next number identified the launch site, with 1 indicating the Kennedy Space Center. Prior to the *Challenger* disaster in 1986, a second launch facility was under development at Vandenberg Air Force Base in California, which would have become launch site no. 2, but construction was subsequently abandoned. The letter designation related to the sequence of each flight in a particular fiscal year. The letter 'A' would be given to the first planned launch and – for example – 'H' would be allocated to the eighth flight. If missions were delayed until the following year or unavoidably flew out of sequence, they would still retain their originally designated number. It was a very confusing system that had been introduced by NASA when the numerical mission sequence could not be maintained due to delays, and it remained in place until shuttle *Challenger* was lost on mission STS-51L (fiscal year 1985, launch at Kennedy, twelfth planned flight).

When shuttle missions resumed in September 1988, the former and less confusing system was reinstated, with the 26th mission designated as STS-26. Despite some later flights being launched out of sequence, the numbering system would continue through to the final shuttle mission, STS-135 (*Atlantis*) in 2011.[5]

## Expanding the Shuttle's Capabilities

Despite early but misguided forecasts that shuttle flights would one day become routine and regular, flying at a rate of one mission every two weeks, that timetable rapidly proved unrealistic, with programme resources strained to the limit, hardware and software problems, unworkable training schedules and a critical lack of availability of spare parts just some of the myriad problems NASA was unable to overcome. The space agency's personnel were working to unsustainable timetables, and parts were being cannibalized from one shuttle to another in an attempt to maintain some semblance of schedule. Rather than the 24 flights a year earlier proudly touted by NASA, the fifteen missions that flew after STS-9 took place over a two-year period. While shuttle *Columbia* was undergoing modifications, a third orbiter, *Discovery* (OV-103), joined the fleet, later followed by the fourth and last planned orbiter, *Atlantis* (OV-104). The highlights of those fifteen missions are summarized below.

### STS-41B: 3 February 1984–11 February 1984 (*Challenger*)

This eight-day mission featured a dual spacewalk conducted by mission specialists Robert Stewart and Bruce McCandless II and presented us with one of the most unforgettable and iconic images of the shuttle era as McCandless became the first human satellite, alone and untethered in his jet-powered Manned Manoeuvring Unit, photographed above the blue planet and silhouetted against the black void of space.

### STS-41C: 6 April 1984–13 April 1984 (*Challenger*)

The main goal of this six-day flight was the first space 'service call', involving the retrieval and repair of the disabled 2,270-kilogram (5,000 lb) Solar Max satellite, launched four years earlier. The first attempt at plucking the faulty satellite out of its orbit failed, but the following day, mission specialist Terry Hart managed to grasp the slowly rotating Solar Max using *Challenger*'s robotic arm, and once it was safely secured in the payload bay, spacewalking astronauts George 'Pinky' Nelson and James 'Ox' van Hoften were able to replace a faulty control box and fix a broken instrument before Hart used the robot arm to successfully redeploy the newly repaired satellite back into orbit.

On STS-41B, MS Bruce McCandless became the first human satellite during this untethered flight away from shuttle *Challenger*.

### STS-41D: 30 August 1984–5 September 1984 (*Discovery*)

The space shuttle *Discovery*'s maiden flight into space was delayed on 25 June 1984, when a faulty computer had to be replaced. The following day, and just four seconds from lift-off, one of the shuttle's main engines began to fire and then shut down when its main fuel line valve suddenly faltered and closed. Two months later, following further frustrating delays, mission 41D finally began with a perfect lift-off on 30 August. One prominent member of the six-person crew was payload specialist Charles Walker, an engineer with McDonnell Douglas, who became the first commercial PS to fly into orbit. He was tasked with monitoring an electrophoresis experiment for his company in the gravity-free environment of space. *Discovery* landed at Edwards on 5 September after a successful six-day mission.

**STS-41G: 5 October 1984–13 October 1984 (*Challenger*)**

On this, *Challenger*'s sixth mission into space, mission specialist Kathy Sullivan became the first American woman to complete an EVA when she and crewmate David Leestma spent three and a half hours floating tethered in the rear of the payload bay, where two fuel tanks were mounted, testing tools that future crews might use in refuelling spent satellites. Unfortunately, she had been beaten to the honour of first woman to perform an EVA by Soviet cosmonaut Svetlana Savitskaya just three months earlier. As well as Sally Ride completing her second space flight, Australian-born oceanographer Paul Scully-Power was on board *Challenger* to monitor and photograph the world's oceans, their currents and spiral eddies.

**STS-51A: 8 November 1984–16 November 1984 (*Discovery*)**

The main objectives for this flight were the launching of two satellites and the attempted retrieval and return to Earth of two faulty satellites that had been left in useless low Earth orbits earlier that year when their rockets misfired. The two satellites carried into orbit – Canada's Anik, and Syncom for the Hughes Space and Communications company – were both successfully deployed, following which, over the next three days, mission specialists Joe Allen and Dale Gardner were able to retrieve both stranded satellites, Palapa-B2 and Weststar-6, and successfully and delicately manoeuvre them into the now empty payload bay for the

MS Dale Gardner holds up a 'For Sale' sign, a cheeky reference to the successful retrieval of the malfunctioning Palapa B2 and Westar 6 satellites.

return journey and refurbishment back on Earth. Controlling the remote arm during this intricate operation was MS Anna Fisher, who became the fourth American woman and the first mother ever to have flown into space.

### STS-51C: 24 January 1985–27 January 1985 (*Discovery*)

Complete secrecy shrouded the next shuttle mission, the first ever dedicated to military use only. No information would be revealed about the large satellite held in the payload bay. Military officials ordered the closure of all Kennedy Space Center press facilities, the withholding of the launch time, and a refusal to broadcast transmissions between ground controllers and the astronauts on board. For the first time, a military payload specialist was part of the crew as part of the Air Force's Manned Spaceflight Engineer Program (MSE), Colonel Gary Payton. As the mission was (and remains) classified, very few details of the three-day flight have ever been revealed.

### STS-51D: 12 April 1985–18 April 1985 (*Discovery*)

McDonnell Douglas engineer Charles Walker would make his second flight into space, operating a large electrophoresis machine – used to isolate molecules – that the company expected would yield a particular compound, the hormone erythropoietin, in sufficient ultra-pure quantities to begin clinical tests for a life-changing drug, in a joint venture between McDonnell Douglas and the Ortho Pharmaceutical division of Johnson & Johnson. If later manufactured, there was hope it could improve life greatly for sufferers of such afflictions as anaemia, and patients needing blood transfusions. It was already proven that weightlessness would allow creating a purity level four times greater than what could be achieved in a ground laboratory. A great deal of other beneficial science was carried out on this mission, but far more media attention was focused on a second payload specialist, in this case U.S. senator Jake Garn. The senator was assigned to the mission due to his chairmanship of a congressional subcommittee that oversaw the NASA budget, and would undergo medical tests during the flight. Unhappily, he suffered badly from space sickness for much of the mission. The crew deployed two communications satellites, Telesat-1 (Anik C-1) and Syncom IV-3 (Leasat-3) into orbit, but the engine on Leasat-3 failed to activate. Using the remote arm, MS Rhea Seddon attempted to prod a toothpick-sized ignition switch but was unable to achieve a result. A spacewalk out to the satellite was considered too hazardous, so MSS Jeff Hoffman and David Griggs performed an unscheduled spacewalk, attaching two 'fly-swatter' tools to the end of the remote arm. Seddon then tried again, using these to flick the switch on Leasat-3, but the attempt failed. Acting on a message relayed from Mission Control, the crew reluctantly abandoned the unresponsive satellite and *Discovery* was then eased away. All of these unsuccessful operations added two extra days to the mission timetable. *Discovery*'s touchdown at the Kennedy Space Center was badly affected by sudden crosswinds, which blew the orbiter 6 m (19 ft) left of the centre of the runway. Karol Bobko brought *Discovery* back under control but had to brake hard, causing the inboard right brakes to lock up, and the inboard tyre blew, causing damage to a wing control flap.[6]

**STS-51B: 29 April 1985–6 May 1985 (*Challenger*)**
When *Challenger* lifted off on the seventeenth space shuttle mission, it was carrying a crew of seven on the second flight of the reusable ESA $1 billion Spacelab 3 laboratory, secured once again within the payload bay and now in its fully operational configuration. It would provide a top-quality microgravity environment for delicate materials processing as well as fluid experiments. It also carried into orbit a small menagerie of animals – two squirrel monkeys and 24 rats – for studies into the effects of weightlessness. On this flight, the seven-man crew was split into Gold and Silver shifts, each working twelve-hour days. After seven days and completing 110 Earth orbits, mission commander Bob Overmyer and pilot Fred Gregory brought *Challenger* down to a smooth touchdown at Edwards Air Force Base.

**STS-51G: 17 June 1985–24 June 1985 (*Discovery*)**
Two foreign guest payload specialists were among the seven crew members assigned to this mission, which would launch and recover the Spartan 1 X-ray observatory and three communications satellites, and carry out a laser-tracking experiment as part of the U.S. military's Strategic Defense Initiative (SDI, a programme also nicknamed 'Star Wars' after the popular science fiction films). Patrick Baudry was a French astronaut who had earlier trained in the Soviet Union as a backup to fellow Frenchman Jean-Loup Chrétien, who had launched aboard a Soviet Soyuz spacecraft in 1982. The second foreign crew member was Prince Sultan Salman Al-Saud, nephew of then king Fahd

The international STS-51G. At rear, MS Shannon Lucid, MS Steve Nagel, MS John Fabian, PS Sultan Salman Al-Saud (Saudi Arabia) and PS Patrick Baudry (France). At front, commander Dan Brandenstein and pilot John Creighton.

of Saudi Arabia, the first member of a royal family, the first Arab and the first Muslim to fly into space. His presence on the crew was essentially a gesture of international goodwill in that he had few responsibilities other than to observe the launch of the Arabsat relay station and to photograph his homeland for geological studies. The 'Star Wars' test involved the installation of a 20-centimetre-wide (8 in.) reflector in *Discovery*'s airlock hatch window that would serve as a target for a low-energy laser in Maui, Hawaii, in a test of high-precision laser-tracking techniques, part of efforts to perfect technologies applicable to the Strategic Defense Initiative. The first attempt failed because of a computer pointing *Discovery* in the wrong direction, but a second, two days later, was successful. Orbiter *Discovery* landed at Edwards Air Force Base after its eighteenth mission, called the 'most successful' to that time, lasting seven days.[7]

**STS-51F: 29 July 1985–6 August 1985 (*Challenger*)**
Described as the most ambitious scientific mission yet flown by any government (and the 50th crewed U.S. space flight), STS-51F had a shaky

## Space Shuttle Missions Scheduled for 1986

| Date | Orbiter | Mission Highlights |
|------|---------|--------------------|
| 12 January | *Columbia* | Launch communications satellite; photograph Halley's Comet; Rep. Bill Nelson on board |
| 23 January | *Challenger* | Launch communications satellite; launch and retrieve small satellite studying Halley's Comet; teacher Christa McAuliffe on board |
| 6 March | *Columbia* | Carry four telescopes to study Halley's Comet, planets, stars, stellar objects, launch satellite |
| 15 May | *Challenger* | Carry Ulysses robotic probe for launch that would take it around Jupiter into orbit of the Sun |
| 20 May | *Atlantis* | Carry Galileo probe for rendezvous with asteroid Amphitrite in December, then on to Jupiter and its moons in 1968 |
| 24 June | *Columbia* | Launch two satellites; British and Indonesian astronauts on board |
| 22 July | *Challenger* | Launch two satellites; Sally Ride on her third flight into space |
| Mid-July | *Discovery* | First shuttle launch from Vandenberg AFB; military mission researching Strategic Defense Initiative equipment |
| 18 August | *Atlantis* | Earth sciences mission |
| 4 September | *Columbia* | Secret military mission |
| 27 September | *Challenger* | Launch satellite for India; first Indian astronaut; possibly first journalist in space |
| 29 September | *Discovery* | Secret military mission from Vandenberg AFB |
| 27 October | *Atlantis* | Launch Hubble Space Telescope |
| 6 November | *Columbia* | Launch two satellites |
| 8 December | *Challenger* | Secret military mission |

start after lift-off when one of *Challenger*'s main engines failed 5 minutes and 40 seconds into the ascent to orbit, fortunately following separation of the solid rocket boosters. Ground controllers were seriously considering ordering a forced emergency landing in Spain, but under the power of its two remaining engines the orbiter was able to limp into a circular orbit 273 km (170 mi.) above Earth – some 116 km (72 mi.) short of its intended target. In order to achieve even this altitude, commander Gordon Fullerton had expended a third of the shuttle's Orbiting Maneuvering System (OMS) and reaction control system (RCS) propellants, which meant they would now not be able to climb to a higher orbit. With the newly configured Spacelab 2 module anchored in its payload bay, this flight was planned as an orbiting astronomical observatory with more than $72 million worth of telescopes and other sensitive instruments to study the Sun, stars and distant galaxies. The lower-than-expected orbit meant that not all of the mission objectives were achieved, although an extra day in orbit did mean that the flight, following a landing at Edwards, was mostly regarded as a success.

### STS-51I: 27 August 1985–3 September 1985 (*Discovery*)

Already three days behind schedule due to bad weather and a computer problem, *Discovery* was finally launched at dawn, in the midst of some of the worst weather seen in the shuttle programme, leaving the launch pad just ahead of an advancing squall. Minutes after lift-off, heavy rain drenched the pad. Ahead of the five crew members was a busy first day in which they would deploy two satellites into orbit. The Australian communications satellite Aussat was the first to be launched into its own orbit, followed just five hours later by a relay satellite owned by the American Satellite Company. A third satellite, Syncom IV-4, would be launched two days later. The crew's next task was the retrieval and repair of the Leasat 3 satellite, the subject of an unsuccessful repair during the STS-51D mission four months earlier. Now equipped with the right tools, mission specialists James van Hoften and William Fisher were able to capture the malfunctioning satellite and coax it into *Discovery*'s payload bay. After two lengthy spacewalks totalling 11 hours and 46 minutes, repairs had been successfully completed and the once-crippled Leasat 3 was released back into orbit. The mission ended after 111 orbits with a safe touchdown on Runway 23 at Edwards.[8]

### STS-51J: 3 October 1985–7 October 1985 (*Atlantis*)

Making a low-key debut, NASA's fourth operational space shuttle, *Atlantis* (OV-104), lifted off from the Kennedy Space Center on a four-day classified military mission for the Department of Defense (DoD), with NASA officials following strict rules imposed by the Pentagon, as with the earlier STS-51C. This meant that very few details about the flight would ever be released. Even the launch time was not revealed until nine minutes before lift-off. Four of the five crew members were NASA astronauts, although it was required that they all have military backgrounds, while the fifth was a second military payload engineer, Air Force Major William Pailes. It is known that two DoD communications systems satellites were launched ahead of the mission-end landing on Runway 23 at Edwards.

**STS-61A: 30 October 1985–6 November 1985 (*Challenger*)**
When *Challenger* lifted off on the orbiter's ninth mission, and the 21st in the space shuttle programme, it not only carried the largest crew (eight) sent into space to that time but was the first mission partially managed by another nation. West Germany's space agency had leased *Challenger* for the eight-day mission, which not only took Spacelab 2 back into orbit but carried three European physicists – Reinhard Furrer, Ernst Messerschmid and Wubbo Ockels. Known as the Spacelab D-1 (Deutschland-1) mission, it was the first to have German mission management and was controlled from the German Space Operations Centre of the German Test and Research Institute for Aviation and Space Flight (DFVLR, a precursor of the present-day German Aerospace Center, or DLR). During the flight more than 75 scientific experiments were carried out in the areas of physiological sciences, materials science, biology and navigation. At the end of the Spacelab D-1 mission, the record crew had completed 110 orbits and logged over 168 hours in space. Sadly, it would prove to be the last landing for the *Challenger* orbiter.[9]

**STS-61B: 27 November 1985–3 December 1985 (*Atlantis*)**
Following a spectacular night launch from the Kennedy Space Center's Launch Complex 39A, the seven astronauts aboard *Atlantis*, now making its second voyage into space, settled in for a week in space during which they would release three commercial communications satellites. During the flight, mission specialists Jerry Ross and Sherwood Spring would perform some high-flying construction work during two spacewalks totalling twelve hours in *Atlantis*'s payload bay, constructing a 14-metre (45 ft) tubular tower out of 99 aluminium struts and later building pyramid-shaped structures 4 m (12 ft) across, to test the human ability to assemble structures in space.

The two payload specialist members of the seven-person crew were the first Mexican in space, Rodolfo Neri Vela, and Charles Walker of McDonnell Douglas, making his third flight. He was on board to continue his work with an electrophoresis device carried in the shuttle's lower deck that utilized weightlessness to generate ultra-pure samples of the erythropoietin hormone. Following a mission lasting just under seven days, *Atlantis* landed one orbit earlier than planned because of lighting concerns at Edwards; but due to the subsequent *Challenger* tragedy, *Atlantis* would not fly again until STS-27, launched in December 1988, three years later.

### STS-61C: 12 January 1986–18 January 1986 (*Columbia*)

Under the command of Robert 'Hoot' Gibson, space shuttle *Columbia* lifted off on a mission to launch the world's most powerful communications satellite, the Satcom KU-1, for the Radio Corporation of America (RCA). Also on board were two NASA mission specialists, astrophysicists Steve Hawley and George 'Pinky' Nelson, who would study and photograph Halley's Comet. Another, and controversial, member of the crew was Democrat Representative Bill Nelson from Florida, chairman of the House subcommittee that oversaw NASA's budget. After a successful mission lasting just over six days, Gibson and his co-pilot Charles Bolden brought *Columbia* down to a landing at Edwards on 18 January, having been diverted from a landing in Florida due to bad weather. It marked the end of the 24th flight in the shuttle programme and the first of what NASA was planning to be a record fifteen missions for 1986. The next mission, STS-51L (*Challenger*), was then scheduled for lift-off on 23 January 1986.

## The Challenger Tragedy

At 11:38 a.m. EST on the morning of Tuesday, 28 January 1986, thousands of excited spectators shouted and cheered with unbridled joy as space shuttle *Challenger* lifted off from the Kennedy Space Center launch pad, cleared the launch tower and climbed skywards on a pillar of white-hot flame. The orbiter automatically executed a single-axis turn at the sixteen-second mark, then slowly arched backwards to assume the correct course to achieve orbit. Everything seemed to be going well on the STS-51L mission as the seven crew members prepared for their passage through the period of highest air turbulence, known as max q, or maximum dynamic pressure.

At this point the orbiter's acceleration was slowed to avoid any possible damage to the shuttle assembly, with the engines throttled back to 65 per cent of full power. Despite this, the crew were still jolted around until the air began to thin around *Challenger*. As the outside pressure decreased, the shuttle's main engines could be restored to full thrust. Following established procedure, mission commander Dick Scobee

*Challenger* lifts off on her final, fatal mission, STS-51L.

transmitted the words 'Throttling up' to fellow astronaut Dick Covey, who was acting as CapCom that day in Mission Control. Covey could see from the data on his screen that the launch was proceeding normally and *Challenger* was following a near-perfect trajectory. He pressed his transmit button and said, '*Challenger*, go at throttle up.' Scobee responded with, 'Roger, go at throttle up!' It was 70 seconds into the flight.

Then, without warning, the shuttle began to shudder and sway alarmingly. Pilot Mike Smith must have realized they were in deep trouble and just had time to exclaim, 'Uh oh!' before a huge fireball flashed along the shuttle stack and the fuel tank erupted in a massive explosion 73 seconds into *Challenger*'s ascent.

The cause of the explosion would later be narrowed down to the failure of an O-ring inside the right-hand solid rocket booster (SRB), due to the unusually cold weather experienced that morning. The frozen O-ring had remained solid, which allowed ferociously hot combustion gases to penetrate through the side of the booster. Photographs later revealed a tell-tale puff of black smoke issuing from a joint in the lower section of the booster at the time of ignition. Even if controllers and sensors had noticed a problem at that time, they would have had no way of shutting down the boosters. From that instant on they were committed to the launch, and the 51L mission was doomed.

Outwardly, everything had seemed to be problem-free as *Challenger* rocketed into the blue Florida sky. What could not be seen with the naked eye was a white-hot flame flaring out from the breach in the right-hand booster, which rapidly grew in intensity. It then began to burn through a strut that held the booster to the huge fuel tank. As the strut melted and disintegrated, the fuel tank to begin to swivel. There was a sudden flash of flame, and the tank erupted in a massive ball of fire.

On the ground, the spectators saw a billowing white, orange and red cloud suddenly envelop the shuttle stack, before two wildly firing boosters emerged from the massive conflagration. Debris showered outwards, trailing white smoke. Nobody could comprehend what they were seeing, and disbelief was quickly followed by the realization that they were witnessing an unspeakable tragedy.

Included in the millions of people across American who had tuned in to the launch were many thousands of school students. Ever since

A massive explosion erupted 73 seconds into the ascent.

a young teacher named Christa McAuliffe had been selected to fly into space and been assigned to the STS-51L crew, students across the United States had been following her progress while she was in training for her mission. Christa had planned to give a series of science lessons from space, and there was great anticipation that students could try to replicate some of her experiments on the ground. Now those children and their shocked teachers knew something had gone terribly wrong. Then came the dreaded announcement that a 'major malfunction' had occurred, and tears began to flow for the beloved young teacher and the rest of her *Challenger* crew.

Christa's parents, Ed and Grace Corrigan, had opted to watch the launch from a crowded car park below the Kennedy Space Center press site. Totally bewildered and shocked by what they had just witnessed, they found themselves being gently escorted by officials to NASA's astronaut crew quarters. It was there that George Abbey, the director of flight crew operations at the time, finally gave them the terrible news that the shuttle had been lost and there was no hope of survival for the crew.

There has always been a misconception that the shuttle itself had exploded, but this was not the case. The force of the wind acting on the shuttle stack while climbing through the atmosphere at such high speed was colossal, and an ascending shuttle has to be pointed with great accuracy. When the solid rocket booster swung loose after the strut had burned through, *Challenger* began to swivel sideways and within moments was literally torn apart. The wreckage of the reinforced crew cabin was sent spiralling out of the conflagration, later to crash into the ocean at high speed.

Two days after the *Challenger* tragedy, President Ronald Reagan gave a moving testimony at a memorial service for the seven astronauts at the Johnson Space Center's central mall. In his speech, he recalled the moving words of John Gillespie Magee's 1941 poem 'High Flight' and reminded those present that the spirit of the American nation was based on heroism and noble sacrifice.

An immediate salvage and recovery operation was mounted by the U.S. Navy following the loss of *Challenger* and her crew. Eventually, pieces of wreckage of all sizes were located and taken aboard salvage ships for transportation back to the Kennedy Space Center, where the highly difficult and complex task of reconstructing the shattered remains was begun, while investigators tried to determine the cause of the tragedy.

The *Challenger* crew: at back, MS Ellison Onizuka, teacher in space Christa McAuliffe, PS Greg Jarvis and MS Judy Resnik. Front row: pilot Michael Smith, mission commander Richard (Dick) Scobee and MS Ron McNair.

On Friday 7 March, Navy divers located the crushed wreckage of the crew module on the ocean floor. The families of the crew were notified that the bodies of their loved ones had been found and recovered, and each of them then began the awful process of arranging a funeral service. Commander Dick Scobee and pilot Mike Smith were buried with full military honours at Arlington National Cemetery, Virginia. Mission specialist Ronald McNair's family held a funeral service in Oregon, while Ellison Onizuka was laid to rest in Hawaii. The ashes of the other two mission specialists, Judy Resnik and Greg Jarvis, were scattered at sea during private ceremonies.

Christa McAuliffe was laid to rest on a hill in Concord's Calvary Cemetery, New Hampshire. The inscription on her gravestone reads: 'S. Christa McAuliffe. Wife. Mother. Teacher. Pioneer Woman. Crew member, Space Shuttle *Challenger*.'[10]

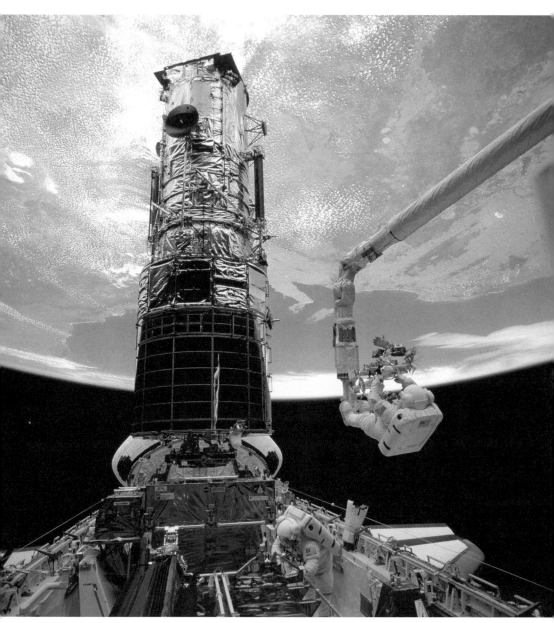

With South Australia as a backdrop, MS Story Musgrave is raised to the top of the
Hubble Space Telescope by Canadarm during the successful STS-61 repair mission.

# 11

# EXPANDING THE SPACE FRONTIER

Thursday, 29 September 1988. It was a day filled with apprehension and even a collective fear. On Launch Complex 39B at the Kennedy Space Center, the same pad from which *Challenger* had left on its final, fateful journey, space shuttle orbiter *Discovery* stood poised, pointing almost defiantly into the blue Florida skies.

Mission STS-26 would be the first shuttle flight since the loss of *Challenger* and her crew 32 months earlier. The previous year, five veteran NASA astronauts had been named to fly the mission: commander Captain Frederick 'Rick' Hauck of the U.S. Navy, pilot Colonel Dick Covey of the Air Force, and three mission specialists, Lieutenant Colonel Dave Hilmers, a Marine, and civilians George 'Pinky' Nelson and Mike Lounge. Now the question everyone was pondering was: how safe would the crew be? Investigators and countless engineers had done their utmost in those two and a half years to create new parameters of safety within the shuttle fleet. NASA had completely overhauled the remaining three orbiters in its fleet, adding hundreds of safety features ordered after the loss of *Challenger*. Morton Thiokol, the Utah-based company that had designed and built the solid rocket boosters, had made many key design changes. But even though confidence was expressed that the problems had been solved, there was still a certain edginess in every reassurance given to the press and public.

The crew of *Discovery* would fly a four-day mission and deploy the $100 million tracking and data relay satellite (TRDS), but essentially STS-26 was a test flight, albeit a crucial one. If anything went seriously wrong, it would place the entire shuttle programme in jeopardy.

On launch day, as thousands of spectators at the spaceport watched, waited and prayed, together with millions viewing the launch live on national television, the lift-off was delayed for 1 hour and 38 minutes due to concerns over winds above 30,000 ft and a few minor technical problems such as faulty fuses that were quickly resolved. Meanwhile, the crew was strapped into the orbiter, wearing – as a safety requirement – bulky partial pressure suits for the first time since the STS-4 mission. As the countdown wound down to the final few seconds there was a slight breeze and an imposing bank of clouds offshore, but not enough to hinder the launch.

At 11:37 a.m. EDT, 975 days after the loss of *Challenger*, her sister ship *Discovery* soared into the sky. Everywhere, anxiety turned to cheers of exultation as the shuttle stack tore a steady path skyward on a trouble-free ascent atop a trail of smoke and orange fire. Then came the moment everyone was dreading. There was a shared silence and not a few shudders across the nation as CapCom John Creighton in Mission Control transmitted the words that sent a chill through everyone listening: '*Discovery*, go at throttle up.' On board the ascending orbiter, commander Rick Hauck calmly stated, 'Roger, go,' and to the relief of everyone the flight continued problem-free through to orbit.[1]

Four days later, two sonic booms at 9:34 a.m. sounded the arrival of *Discovery* over the desert-based Edwards Air Force Base, where an estimated 425,000 flag-waving people had gathered to witness the triumphant finale to America's return to space. A trail of dust from dry lake Runway 17 swirled upwards as *Discovery* touched down at 332 km/h (206 mph), travelling 2,330 m (7,644 ft) along the hard clay runway before coming to a halt just 49 seconds after landing.

Vice President George H. W. Bush greeted the crew as they disembarked 55 minutes later, congratulating them on a highly successful mission. 'All we can say from our hearts is thank you; thank you for putting us back in space,' he said. 'You've proven that the space program is stronger than ever, and stronger supported than ever.'[2]

### From Strength to Strength

While it is beyond the scope of this book to describe the 87 space shuttle missions that followed the loss of *Challenger* on 26 January 1986 and prior to the loss of *Columbia* on 1 February 2003, there would be numerous highlights. These included the deployment of the massive

Hubble Space Telescope (HST) and subsequent repair and servicing missions; nine shuttle missions to link up with Russia's Mir space station; the deployment in orbit of the *Magellan* Venus, *Galileo* Jupiter and *Ulysses* solar probes from the shuttle's payload bay; working aboard the Astro 1 and 2 observatories and numerous Spacelab missions; NASA's Spacehab module (which could hold spare parts, equipment and supplies, or even act as a laboratory on its own, depending on the mission); seven Department of Defense missions; the first flight of replacement shuttle *Endeavour* (STS-49, launched 7 May 1992); and even the second launch of Mercury astronaut John Glenn into orbit on 29 October 1988 aboard *Discovery* on shuttle mission STS-95. Some facts pertaining to those interim missions are as follows:

★ All 87 launches took place from the Kennedy Space Center
★ Cumulative total of 532 crew members launched into orbit
★ Orbiter *Discovery* flew 24 missions, *Atlantis* 24, *Columbia* twenty and *Endeavour* nineteen
★ 56 missions touched down at Kennedy Space Center; 31 at Edwards Air Force Base
★ Nine shuttle missions to Mir; sixteen assembly/supply missions to the ISS
★ Longest orbital flight: STS-80 (*Columbia*), 17 days, 15 mins
★ Shortest orbital flight: STS-83 (*Columbia*), 3 days, 23 mins (due to a fuel cell problem)
★ Highest altitude reached: STS-82 HST servicing, 619.6 km/ 385 mi.

One flight gained far more attention than any other during this period: STS-61, with the shuttle *Endeavour*, launched on a dangerous but crucial repair mission to the orbiting but imperfect Hubble Space Telescope. It was a dramatic operation that involved four mission specialists conducting several intricate, lengthy and painstaking spacewalks over five days, as the world looked on.

### A Flawed Eye in the Sky

One of the most remarkable scientific facilities ever devised totally revolutionized our understanding of the cosmos, enabling astronomers to undertake studies they could previously only dream about. That facility – for want of a better word – was the Hubble Space Telescope, or HST. The bus-sized HST orbits high above the clouds, but after it had been placed in orbit during the STS-31 mission in 1990, some clouds of a vastly different kind enveloped the giant telescope. To the dismay and ire of NASA, the $1.6 billion telescope was found to be suffering from an unexpected and severe case of myopia. The first blurred images from the orbiting telescope revealed an unexpected and serious manufacturing flaw: a deformity in its primary mirror.

The U.S. Congress had authorized the construction of the HST in 1977, naming the space telescope after Edwin Hubble, the foremost American astronomer of the twentieth century. Following on from this approval, work on the massive telescope was divided among several institutions. Responsibility for the design, development and construction of the HST was handed to NASA's Marshall Space Flight Center (MSFC) in Huntsville, Alabama, while the administration and control of all the telescope's scientific instruments fell to NASA's Goddard Space Flight Center in Greenbelt, Maryland. Optics company Perkin-Elmer in Danbury, Connecticut, was commissioned by MSFC to design and construct the Optical Telescope Assembly (OTA) and fine guidance sensors for the space telescope.[3] Another major contractor company was Lockheed, commissioned to construct and integrate the unique spacecraft inside which the telescope would be housed.

The HST was expected to look deeper into space than ever before during its lifetime, sending back data and images that might help explain many unsolved mysteries of our universe, using a primary mirror 2.4 m (7.9 ft) in diameter to focus light rays emanating from far distant galaxies and stars. If all went as planned, scientific data would be transmitted from HST to a NASA communications satellite, and on to the receiving station at White Sands, New Mexico. Then, a commercial communications satellite would relay the signals to NASA's Goddard Space Center, and finally, land lines would carry them to the Space Telescope Science Institute in Baltimore, Maryland. All the data would finally be passed on to anxiously waiting astronomers.

Work on the HST was finally completed, and its launch and deployment into orbit was tentatively scheduled for a shuttle mission in October 1986. The loss of space shuttle *Challenger* early that year had put an abrupt end to those plans, grounding the shuttle fleet and the space telescope for the foreseeable future. The HST was subsequently housed in a pristine 'clean room', powered up and purged with nitrogen until a new launch date could be set. Some tests and improvements were carried out as the months rolled by, until shuttle flights finally resumed in 1988. The Hubble launch was then allocated to the STS-31 mission, scheduled for 26 March 1990.

Following a series of delays, the HST was launched aboard *Discovery* from Kennedy at 8:33 a.m. EDT on 24 April 1990, embarking on one of the most complex missions to that time. It was successfully deployed into orbit around 600 km (370 mi.) above the planet by *Discovery*'s crew. Within days of the HST being placed into orbit there was deep concern among the telescope's operators that its primary mirror seemed to have an aberration that was badly affecting the sharpness of the telescope's early images. NASA undertook an investigation to diagnose the problem.

According to NASA:

> Ultimately the problem was traced to miscalibrated equipment during the mirror's manufacture. The result was a mirror with an aberration one-50th the thickness of a human hair, in the grinding of the mirror. Replacing the mirror was not practical, so the best solution was to build replacement instruments that fixed the flaw much the same way a pair of glasses correct the vision of a near-sighted person.[4]

Corrective optics and new instruments were built and were to be installed on Hubble by four spacewalking astronauts, Story Musgrave, Jeffrey Hoffman, Kathryn Thornton and Tom Akers, during the STS-61/*Endeavour* shuttle mission in late 1993. It was a highly dangerous, complex rescue mission, and the mission specialists involved in the repair mission worked seventy-hour weeks for ten months, including a total of four hundred hours training underwater in a weightlessness simulation pool, rehearsing every aspect of the mission.

As well as giving HST that new 'pair of glasses', they had to install six new gyroscopes, with which the telescope locks on to designated targets. Three existing ones had failed and there was considerable anxiety that if one more were to fail, the HST might cease working. They also had to replace the telescope's two solar panels, which supplied it with electricity, with two new ones built by British Aerospace. The existing panels had developed a wobble caused by extreme changes of temperature as HST moved from sunlight to darkness and back.

On 2 December 1993, shuttle orbiter *Endeavour* lifted off on what would be one of the most hazardous and intricate missions ever attempted. The seven astronauts, under the steady leadership of Dick Covey, were as prepared as they could be, but they knew that if they failed to repair the HST, or caused any damage during their spacewalks, the reputation of NASA would take a severe and extremely expensive hit.

Once in orbit, Covey and pilot Ken Bowersox set about rendez-vousing with Hubble, allowing the orbiting telescope to be captured by the shuttle's robotic arm and deposited in *Endeavour*'s payload bay, ready for the repair work to begin. Then, from 5 to 9 December, five separate EVAs were carried out, each by two astronauts:

> EVA 1 (5 December): Musgrave and Hoffman, 7 hrs, 54 mins
> EVA 2 (6 December): Thornton and Akers, 6 hrs, 36 mins
> EVA 3 (7 December): Musgrave and Hoffman, 6 hrs, 47 mins
> EVA 4 (8 December): Thornton and Akers, 6 hrs, 50 mins
> EVA 5 (9 December): Musgrave and Hoffman, 7 hrs, 21 mins
>
> Total EVA time: 35 hrs, 28 mins

In accomplishing those five work-intensive spacewalks, the four astro-nauts replaced the telescope's Wide Field and Planetary Camera (WFPC) with the updated WFPC2. They also installed in Hubble a new device, the Corrective Optics Space Telescope Axial Replacement (COSTAR) instrument – roughly the same size as a telephone booth – that would replace Hubble's High Speed Photometer (HSP) in order to correct optical problems in the primary mirror for the remaining instruments. The mission proved to be an unqualified success, and to the delight of the telescope's controllers, the telescope soon began operating at its full potential, returning spectacular images of cosmic phenomena that exceeded even the greatest expectations of astronomers around the world.

There would be three subsequent service missions to Hubble, in 1997, 1999 and 2002 (STS-82, 103 and 109, respectively). During these missions, spacewalking astronauts repaired the telescope's gyroscopes as well as installing new instruments. These included a near-infrared spectrometer and a wide-field camera.

On 11 May 2009, the fifth and final HST mission, STS-125, lifted off under the command of veteran astronaut Scott Altman. The crew of *Atlantis* were able to repair and upgrade the HST, carrying out five spacewalks designed to extend the life of the orbiting telescope. Two new instruments, the Cosmic Origins Spectrograph (cos) and an updated wfpc3 were installed, and two others repaired, while gyroscopes and batteries were replaced. New thermal insulation panels were also installed.

Following a successful thirteen-day mission, *Atlantis* touched down at the Kennedy Space Center on 24 May, having given Hubble an extended operational lifespan. It is expected to remain operational – despite recent issues with some of its instruments – until the mid-2020s, by which time it is expected to have been replaced by the James Webb Space Telescope (jwst). This will be an orbiting infrared observatory designed to complement and extend discoveries made by the Hubble Space Telescope, but with vastly improved sensitivity and far longer wavelength coverage. It will be equipped with a mirror seven times larger than that on the hst.

## The *Columbia* Tragedy

'Lock the doors!' With those three quietly spoken words, nasa flight director LeRoy Cain confirmed the unthinkable – that the United States had lost another space shuttle and crew. America's entire human space flight programme was once again thrown into immediate disarray.

Cain's curt command set in motion a sequence that by now had become standard operating procedure for unforeseen events. All of the data in the mission control centre computers had to be frozen, reports finalized, and immediate accounts written of what everyone had just seen, heard and done. Nobody could leave the room or even make phone calls.

The sense of disbelief was palpable. No one in Mission Control had had any reason to believe that *Columbia*'s re-entry would be any different to those of the previous 111 shuttle missions. Controllers and the public alike had become accustomed to the sight of shuttle orbiters gracefully touching down at mission's end.

This flight, designated sts-107, had begun on 16 January 2003 like so many others, with a lift-off from the Kennedy Space Center, carrying a crew of seven trained to perform a series of microgravity science experiments. Most flight activity at that time was centred on crews building and occupying the International Space Station, and during a $145 million overhaul completed in 2001, *Columbia* had actually received most of the modifications needed to rendezvous and link up with the orbiting station. However, STS-107 had been planned as a purely Earth-orbiting research mission and was not scheduled to dock with the ISS. The orbiter, a futuristic tool built on 1970s technology, was scheduled to fly to and link up with the ISS on a later flight in 2003, but that particular flight of America's first operational space shuttle would never take place.

The crew on *Columbia*'s 28th mission into space comprised mission commander Colonel Rick Husband; pilot Commander William 'Willie' McCool; payload commander and mission specialist Lieutenant Colonel Michael Anderson; payload specialist Ilan Ramon, a colonel in the Israeli Air Force; mission specialist Kalpana Chawla; mission specialist Captain David Brown; and mission specialist Captain Laurel Clark.

At 82 seconds into the launch, a large chunk of foam insulation dislodged and fell from the 'bipod ramp', part of the structure used to attach the external fuel tank to the shuttle. Such foam shedding was not unusual; in fact it was quite a common occurrence in previous lift-offs, even reported as possibly having happened as far back as the first shuttle mission, STS-1. Remarkably, no preventive action had ever been

The STS-107 crew. Back (from left) MS David Brown, MS Laurel Clark, MS Michael Anderson and PS Ilan Ramon (Israel). Front: mission commander Rick Husband, MS Kalpana Chawla and Pilot Willie McCool.

taken. Video footage of the incident was reviewed by engineers, who determined that the piece of foam – about the size of a small suitcase – had actually impacted the leading edge of the shuttle's left wing, but they were unable to ascertain at the time whether it had caused any damage. Furthermore, NASA engineers believed that if extensive damage had occurred, the crew members wouldn't be able to remedy the situation while the shuttle was in orbit.

Nevertheless, a number of people within NASA were concerned enough to request photographs of the wing, as the crew were unable to see that section of the wing and had no EVA capability on board. The Department of Defense was reportedly prepared to use its orbital spy cameras to focus on and check the wing, but NASA declined the offer as too involved and intricate an operation for what they believed was probably a false alarm.

On the morning of 1 February 2003, following the sixteen-day science mission, *Columbia* began its landing approach to the Kennedy Space Center. At 7:53 a.m. local time in Houston, Texas, the shuttle was sweeping supersonically high above San Francisco on its landing approach, bound for the Kennedy Space Center thousands of kilometres away in Florida, when worrying anomalies began showing up at Mission Control. Data on their monitors began to flicker, suggesting a problem with sensor information on the temperature of the hydraulic systems in the veteran orbiter's left wing. Then, the sensors stopped working altogether. The Mission Control centre guidance and navigation officer reported that *Columbia*'s wing was encountering unexpected drag, or wind resistance.

Six minutes later, CapCom and fellow astronaut Charlie Hobaugh contacted the Shuttle's commander, Rick Husband, about rising tyre pressures. 'And *Columbia*, Houston, we see your tyre pressure messages and we did not copy your last.' 'Roger,' responded Husband, 'bu– . . .'. Abruptly, there was static, followed by an awful silence. A feeling of dread began to take hold in the minds of the controllers. It wasn't long before a horrible reality set in, and once again America's space programme had changed forever.

People right across the southern states reported hearing a tremendous boom, followed by a long, low rumbling noise. Some who knew the shuttle would be passing overhead at that time on its descent path were filled with concern. They would soon learn that *Columbia* had been torn to pieces at an altitude of 55,000 m (181,000 ft) a little southeast of Dallas, while hurtling towards Florida at more than 17,700 km/h (11,000 mph). Larger solid parts of the shuttle, such as the main engines, continued to streak brightly across the morning sky for a while, vividly captured in several photographs that would soon hit newspaper offices, while tens of thousands of smaller pieces of debris would rain down over vast tracts of Texas and Louisiana for the next half-hour. Everyone at NASA quickly learned that there had been another catastrophic event, and another shuttle crew had been lost. Shocked crews of future missions would be forced to bide their time and continue training as the painful, lengthy process of self-evaluation and gathering data on this latest calamity got under way.

In August 2003, after carrying out countless tests and interviews, data reconstructions and debris reviews, the Columbia Accident Investigation Board (CAIB) officially found that the primary physical cause in the loss of the shuttle centred on the shedding of a 750-gram (1.6 lb) chunk of foam insulation from a support spar on *Columbia*'s external fuel tank. It had struck and punctured the left wing, leaving a hole in one of the most critical areas of the shuttle's protective surfaces. During re-entry, superheated atmospheric gases seeped through this breach as *Columbia* tore through its fiery re-entry, leading to the loss of the sensors and eventually melting the wing frame, following which the shuttle broke apart. There may have been some worrying indications of a problem for the returning crew, but they would have died almost instantly when *Columbia* broke apart.

'NASA's organizational culture had as much to do with this accident as foam did,' the report concluded.[5] It did not direct all of its criticism at the space agency, placing blame too on the White House and Congress for a steady decline in NASA's budget, with adjustments for inflation. This had created a trend, the report stated, that forced NASA to cut its workforce and rely excessively on outside contractors. Nevertheless, the board was scathing of NASA's work, its organizational barriers, communication and management practices, saying the space agency had done little to improve shuttle safety since the loss of *Challenger*. Without a clear commitment to change, the board iterated, NASA might once again repeat fatal practices that could result in the loss of another shuttle and its crew.

Of its 29 recommendations, the board concluded that before the next shuttle flew, at least fifteen should be implemented, and that NASA should 'develop a practicable capability to inspect and effect emergency repairs to the widest possible range of damage'.[6] An independent panel of experts chaired by former astronauts Tom Stafford and Dick Covey was convened, which closely monitored NASA's implementation of the recommendations.

On 21 December 2001 a new NASA administrator had taken over the agency's reins from incumbent Dan Goldin thirteen months before the loss of *Columbia*. The day before the CAIB report had been released, a disconsolate Sean O'Keefe acknowledged at a press conference that NASA had 'just plain missed' the serious consequences of a large piece of foam striking the wing during the launch ascent. He later declared the report would serve as NASA's blueprint for the future. 'We have accepted the findings,' he said, 'and will comply with the recommendations to the best of our ability.'[7]

The aftermath of the tragedy was far-reaching in many ways, especially for NASA's future plans. Earlier assembly flights had meant the International Space Station was well on the way to completion, although it had become something of a notorious money sink, while at the same time the shuttle fleet was drawing towards the end of its projected useful lifespan. Meanwhile, programmes for a new generation of manned spacecraft and crew rescue vehicles kept falling victim to savage budget

STS-116 mission specialists Robert Curbeam (left) and Christer Fuglesang photographed during EVA work on the construction of the ISS International Truss Structure (ITS), December 2006.

On 28 July 2005, the shuttle *Discovery* performed a 'flip' manoeuvre prior to docking with the ISS. This allowed the crew onboard the ISS to take detailed photographs of the shuttle's thermal protection system and verify its integrity. As mission commander Eileen Collins performed the 'flip', *Discovery* and the ISS were about 200 m (656 ft) apart, at around 353 km (219 mi.) above Switzerland.

cuts, and there seemed to be no clear indication of where America would stand in space exploration in another decade.

As one necessary interim measure, with America's shuttle programme now on hold, three-seat Russian Soyuz vehicles were used to transport cosmonauts, astronauts and other international crew members to and from the ISS. At least one Soyuz craft would always remain docked to the station in case it was ever needed as an emergency evacuation vehicle.

On 14 January 2004, less than a year after the loss of *Columbia*, President George W. Bush breathed a little welcome life into NASA when he announced his bold vision for the agency. This included sending human explorers back to the Moon by 2020, and eventually to Mars. President Bush is then said to have consulted NASA administrator O'Keefe and declared, 'Tell me what you need to fix the agency,' following which formal discussions were arranged. The irony is that these plans only came about as a political reaction to the most recent shuttle disaster.

The external fuel tank and its insulating foam coating underwent a design overhaul, and other safety measures were implemented. Then, on 26 July 2005, STS-114 (*Discovery*) lifted off with a crew trained to test a range of new procedures, which included using cameras and a robotic arm to scan the shuttle's underbelly for any broken or lost tiles. More cameras were installed on the launch pad, to film each shuttle launch and better monitor any further foam shedding. This actually showed that despite preventive measures, the foam loss was still heavier than expected, so the next shuttle flight, STS-121 (also *Discovery*), did not take place for another twelve months. Once this flight had safely concluded, NASA deemed it safe for the programme to continue, and shuttle missions resumed according to the newly revised schedule.[8]

Over three decades from 1981 to 2011, America's space shuttle would eventually launch on a total of 135 missions. Following *Columbia*'s maiden flight in 1981, three more orbiters had followed the veteran shuttle into the Florida skies: *Challenger*, *Discovery* and *Atlantis*. A fifth operational shuttle, *Endeavour*, began orbital operations in 1991 as a replacement for *Challenger*.

On 21 July 2011, shuttle orbiter *Atlantis* touched down on the Kennedy Space Center runway at the end of mission STS-135, following a mission lasting twelve days and a little over eighteen hours. There were four crew members on board on this final flight by *Atlantis*, which also brought to an end the incredible thirty-year history of the space shuttle programme.

The last space shuttle crew, STS-135. From left: MS Rex Walheim, the pilot Doug Hurley, the commander Chris Ferguson and MS Sandy Magnus.

NASA did not have the sole discretion to terminate the space shuttle progamme; NASA is a government agency, and it was the U.S. government that was ultimately responsible for the decision. Questions had been raised; each of the shuttles had been designed for an operating lifetime of one hundred flights; the three remaining shuttles had flown just 25, 33 and 39 times respectively. So why was the programme cancelled? There were several factors. Although the shuttle had been crucial in the construction of the International Space Station, it was seen to have no place in the next phase of space exploration, the programme known as Constellation – later cancelled – which planned to send astronauts beyond low Earth orbit. In addition, the shuttle programme was massively expensive to operate within an increasingly tight budget, a budget that included ongoing maintenance and upgrading of three workhorse shuttles that were more than twenty years old. A decision had to be made, as the shuttle and the ISS were dominating NASA's budget, to the

*Atlantis* takes off for the final time on 8 July 2011. STS-135 marked the last flight for America's fleet of space shuttles.

*Atlantis* touches down for a night landing on 21 July 2011, ending America's space shuttle programme.

point where the space agency was unable to advance to the next stage of human space exploration. It was decided that either the ISS or the space shuttle – which had never lived up to its earlier promise of low-cost and routine access to space – had to go. There was still much important research work to be carried out during the extended lifetime of the ISS, so it was reluctantly decided to retire the space shuttle fleet and for NASA to devote its budgetary resources to sending astronauts beyond low Earth orbit, a return to the Moon and eventually on missions to Mars. For much the same reason, the ISS will one day face retirement and probable deorbiting. Dates for that retirement, in a few years, are already under discussion.

### Building the International Space Station

As early as 1968, NASA planners had begun looking to a future well beyond the Apollo lunar programme, with their focus centred on an Earth-orbiting station – the next logical step in human space exploration. With long-duration science missions a major objective, they needed to find some means of resupplying the station, and it was determined that a supply craft needed to be developed that would shuttle between Earth and the orbiting station. Thus, in the summer of 1968, the word 'shuttle' first began to appear in NASA's technical literature.

There was agreement that the space station would have to be mostly constructed in orbit, with modules being transported and added to a large central core. The cost, however, was prohibitive, as the only dependable heavy transportation system then available was the Saturn rocket. It was an expendable rocket, and hugely expensive at a time when NASA was beginning to suffer the effects of massive budget cuts, so consideration had to be given to the foresight involved in the design and purpose of this new shuttle vehicle.

By 1972 President Nixon had granted approval for the development of a reusable space transportation system, although it was not linked to any long-term strategy. The proposed vehicle was given the official title of 'space shuttle', and it would be developed first, as NASA was already in the process of preparing to launch its Skylab space station and three visiting science crews, using the remaining Saturn rockets. The production of more of these massive rockets had already been cancelled as a cost-cutting measure. There would be no need for a resupply system as the crews would have adequate provisions on board the station to cover their short-duration missions.

In February 1974 the third and final three-man crew abandoned the Skylab space station. Only one Saturn rocket remained, and this had already been allocated to the Apollo–Soyuz Test Project flight the following year. It was a frustration for NASA and the astronauts, as there was anticipation that Skylab would remain in an ever-decreasing orbit until 1983. With an expectation that the space shuttle could be flying in 1979, contingency plans were put in place in 1977 for the second crewed shuttle test flight to rendezvous with the space station. The crew would then install on the station a small rocket device known as the Skylab Reboost Module, which would fire and insert Skylab into a far higher orbit for the possible reintroduction of crew habitation missions. These plans were thwarted by the development of the space shuttle taking far longer than expected and some higher-than-expected solar activity causing a significant retardation in the station's orbit. The abandoned station re-entered Earth's atmosphere in July 1979 and was mostly incinerated over the west Australian desert.[9]

Following the first crewed space shuttle orbital test flight in April 1981, plans for a modular space station advanced with the establishment in May the following year of a Space Station Task Force. Their concept and recommendations were later approved, and in January 1984 President Ronald Reagan announced that he was committing the United States to the development of a permanently manned space station within a decade. In April that year the Space Station Program Office was established, and twelve months later, eight contractors had been assigned to the task of expanding the station concept into detailed plans and submissions.

During his 1984 State of the Union address, President Reagan invited other countries to participate in the development, construction and occupancy of the proposed space station. Canada, Japan and the thirteen nations that made up the ESA agreed to come on board, similarly devoted to the peaceful uses of space. Greatly easing some of the potential design, cost and construction issues, the European and Japanese space agencies enabled the United States to reduce the number of American-made laboratory modules on the space station by allowing their new partners to provide one module each. The design now emerging was one of a station with a dual-keel configuration and a central truss incorporating the main living and working quarters, in addition to panels of solar arrays.

The station's design phase ended in January 1987 and the first development contracts were awarded and announced that December. On 29 September 1988 the four parties formally agreed to name the space station *Freedom*.

The project, however, was continually beset with funding and technical problems, causing lengthy delays. The original estimated cost of the space station was in the vicinity of $8 billion, but by 1993 this figure had soared to $38.3 billion, which then President Bill Clinton considered to be unsustainable. He therefore called for a complete redesign of the space station to significantly lower the cost, and called on NASA to expand participation by inviting even more international partners. Russia was approached by NASA administrator Daniel Goldin and, following high-level discussions, they agreed to merge their separate space station plans into a single facility. With the Russians now on board, space station *Freedom* underwent a White House-driven name change to the rather more diplomatic name of *Alpha*. By agreement, *Alpha* would utilize 75 per cent of the hardware from *Freedom*, while Russia would contribute parts of its unflown Mir-2 space station to assist in lowering the overall cost. In the course of these negotiations, the mundane moniker *Alpha* was also discarded, replaced by the non-controversial name of International Space Station, or ISS. The Johnson Space Center in Houston would become the lead NASA centre behind

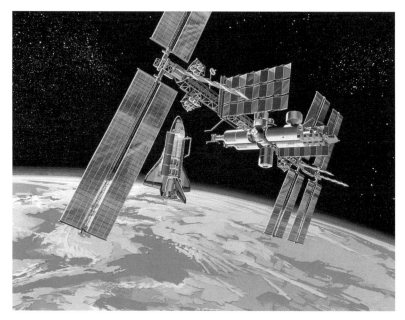

An early NASA design of the space station *Freedom*.

the ISS programme, while in August 1993 the space agency selected Boeing as prime contractor for the space station. Boeing would be responsible for the design, development and integration of the ISS and the construction of all of the major U.S. elements of the station, as well as the Space Station Processing Facility at the Kennedy Space Center in Florida.[10]

From 1998 on, the International Space Station took around ten years and more than thirty missions to assemble, resulting from unprecedented scientific and engineering collaboration among five space agencies representing fifteen countries. Today, orbiting 400 km (250 mi.) above our planet, it is five times larger than the U.S. Skylab station.

The first modular element of the ISS, known as the functional control block, was called *Zarya* (Sunrise), launched aboard a Russian Proton rocket on 20 November 1998. This was followed two weeks later on 4 December by *Unity*, the first of three eventual American node modules carried aboard space shuttle *Endeavour* on the STS-88 mission, linking up to and connecting in orbit with *Zarya*. In July 2000 another Russian module, *Zvezda* (Star), was launched and connected. It included such vital components as the flight-control systems, life support and living quarters, and could support up to six crew members, including separate sleeping quarters for two persons at a time. On 2 November that year the ISS received its first resident crew: Russian cosmonauts Sergei Krikalev and Yuri Gidzenko and NASA astronaut William Shepherd, who flew up in a Soyuz spacecraft." Since then, the ISS has been continuously occupied by literally hundreds of astronauts and cosmonauts on what are known as progressive expedition crews.

NASA's microgravity laboratory *Destiny* and other elements were subsequently added to the growing station, ferried up by a succession of shuttle and Soyuz missions and connected in an ongoing series of spacewalks. In addition to crewed ferry flights, Russian modules were launched into orbit atop expendable Russian rockets, rendezvousing and docking with the ISS under automatic control. Over the next few years a vast complex of laboratories and habitats evolved, crossed by a lengthy truss supporting four units that held large solar-power arrays and thermal radiators. As well as the United States and Russia, the ongoing construction of the ISS also involved the assistance of Canada, Japan, Brazil and eleven ESA members. While space shuttles and Soyuz spacecraft transported crews to and from the ISS, a rotation of Soyuz craft would always remain docked to the space station, serving as a 'life-boat' in the event of an emergency requiring an immediate evacuation.

Following the loss of *Columbia* on 1 February 2003 (on a non-ISS science mission), NASA's space shuttle programme was halted, which impacted for a considerable time on the expansion of the ISS. During this period crews flew up to and returned from the ISS aboard Soyuz spacecraft, while the station was serviced by a series of Russian auto-mated cargo ferries named *Progress*.

The final chapter in the thirty-year history of space shuttle missions began on 8 July 2011 with the launch of a four-person crew aboard orbiter *Atlantis* on mission STS-135. Two days later the crew docked with the ISS, the link-up announced on the ISS by the ceremonial ringing of the station's bell, welcoming a visiting space shuttle for the final time. On 19 July, *Atlantis* undocked from the ISS for a successful but poignant journey home, ending NASA's space shuttle programme. From that time on, American astronauts could only journey to the ISS aboard Russian Soyuz spacecraft.

Today, the ISS is a relatively comfortable six-bedroom research station featuring two bathrooms, a gym, and a cupola providing exten-sive panoramic views of Earth below. The station's power is provided by an acre of solar panels, which make it particularly easy to spot from the ground as it crosses the evening sky.

The ISS as photographed by the crew of STS-132 on 23 May 2010.

Over the more than two decades of its existence, the ISS, continuously occupied since November 2000, has played a unique and critical role as an orbiting science and medical laboratory and observatory. To February 2021, a total of 64 expedition crews, comprising 240 individuals from nineteen different countries (a number which includes 34 women and eight 'space tourists'), have visited the ISS, with an abundance of precious data returned from many long-duration missions.

In some of its more notable records, U.S. astronaut Peggy Whitson holds the record for the total time living and working in space, at 665 days (over three missions), while Scott Kelly holds the long-duration record for most days on the ISS during a single mission, at 340 days. A third U.S. astronaut, Christina Koch, holds the female record for a single-mission residency of the ISS, at 328 days.

### Buran: The Soviet Space Shuttle

To a nation now familiar with the spectacular sight of a delta-winged NASA space shuttle perched on the launch pad at the Kennedy Space Center, images later seen from half a world away in Soviet Central Asia bore many remarkable similarities. The date in question was 15 November 1988 and a near-identical, 100-tonne (110 ton) delta-winged spacecraft, of which 14 tonnes (15.4 tons) was fuel, stood majestically on the launch pad in the bright glare of spotlights at the Baikonur Cosmodrome, southern Kazakhstan. Unlike every U.S. shuttle launch, however, there was no human presence on board – the Soviet shuttle would be unmanned throughout its orbital mission.

Mounted onto a non-recoverable Energia rocket, the Soviet shuttle was known as *Buran* (Snowstorm). Outwardly, it closely resembled America's space shuttle. The *Buran* orbiter stood 36.4 m (120 ft) tall with a wingspan of around 24 m (79 ft). Like its American counterpart, it had around 38,000 individually crafted thermal protective tiles cemented to the aluminium exterior of the vehicle. The main propulsion system at the rear of the fuselage comprised two groups of manoeuvring rockets, with another group situated at the front of the orbiter. Unlike the American shuttle, *Buran* did not feature any main engines and was little more than an inert payload of the Energia booster, rather than functioning as an integral part of the launch process. The lack of main engines meant that the *Buran* could carry a much larger payload into space than the NASA orbiters, as well as having its landing weight similarly increased.

Another notable difference was that there was provision for the future inclusion of two jet engines on the rear of *Buran*, an extra safety factor that allowed for the final stage of landing to be carried out under power. This would allow the cosmonaut pilots to accept a wave-off and go-around in the event of adverse circumstances, such as strong crosswinds over the runway. By way of contrast, America's shuttle fleet was designed to complete unpowered glider landings, without any chance of making a second approach. Despite the concept being deemed workable on crew-tended *Buran* missions, it would never be implemented.

At the scheduled time in the darkness of early morning, the Energia's four first-stage engines roared into lusty life. Slowly at first, the rocket lifted off, rapidly accelerating and carrying the unmanned *Buran* aloft, flying a flawless trajectory into space. Following the booster separation phase, *Buran* continued on its path into orbit.

Energia–*Buran* stands ready for launch on Pad 110 at the Baikonur Cosmodrome.

After just two planned orbits of Earth, *Buran*'s retrorockets automatically fired as it passed backwards over Chile before swinging around in preparation for the fiery re-entry. The first flight of a Buran spacecraft ended right on schedule with a successful and fully automatic touchdown on a specially constructed concrete runway 5 km (3 mi.) long and 80 m (262 ft) wide, situated just 12 km (7.5 mi.) from the Baikonur launch pad. The landing was almost flawless, despite a strong crosswind estimated at around 60 km/h (37 mph) – so accurate, in fact, that the nose wheels only missed the runway's centre line by a mere 1.3 m (4 ft). The flight, which eventually proved to be both the first and last flight of a *Buran* shuttle, had lasted a total of 205 minutes.

Although the flight was hailed as a great triumph of Soviet engineering and technology, the delta-winged craft had sustained a considerable amount of heat damage during re-entry. Soviet authorities decided not to repair the damage immediately as it would prove extremely costly and there was no money available. The following year fresh plans were tabled, in which the *Buran* might be launched on a second orbital mission in 1993 – once again without the benefit of a crew – and a manifested flight duration of some fifteen to twenty days. Those plans, however, were eventually shelved as impractical (and again unsustainably expensive), and the *Buran* was retired after just one mission into space.

With the demise of the Soviet Union towards the end of 1991, the immense cost associated with maintaining the Energia-*Buran* programme caused its demise. In 1993, after eighteen years of development, it ended. The flown *Buran* had meanwhile become the property of the government of Kazakhstan and was housed in Hangar 112 at the Baikonur Cosmodrome, perched on top of an Energia rocket mock-up. Due to the extreme heat of summer and bitter cold of winter in that region, the hangar was thermally insulated with a thick layer of foam,

*Buran* touches down at the end of its two-orbit first and last flight.

but when it rained the foam soaked up any intruding moisture and became increasingly heavy. On 12 May 2002, torrential rain struck the area and the roof became so waterlogged that it finally collapsed. Three out of five roof segments crashed 80 m (262 ft) to the ground, destroying the historic spacecraft and killing eight workmen engaged in trying to repair the roof.[12]

### Astronauts, Cosmonauts, now Taikonauts

In September 1992 the Chinese Manned Space agency (CMS) formally came into operation, together with what became known as Project 921, the nation's three-step thrust for placing Chinese nationals in space and launching China's own permanent space station. Research would be carried out aboard the station by Chinese space travellers, known in this instance as 'taikonauts'. The term is derived from the Mandarin Chinese word *taikong* (space), along with the '-naut' suffix used by U.S. and Russian spacemen, which in turn derives from the ancient Greek for 'sailor' and translates to 'traveller'. Thus 'taikonaut' simply means 'space traveller'.

The People's Liberation Army Astronaut Corps was established in 1998 and selected its first group of fourteen taikonauts that same year. A second cadre of seven was selected in 2010 to support China's long-term goals for space exploration, which will eventually involve flying crewed missions to the Moon and Mars.

The first step in China's manned space programme took place five years later. Following on from the achievements of the U.S. space programme and that of the Soviet (later Russian) programme, China became the third nation to independently send humans into space. Taikonaut Yang Liwei was launched into Earth orbit aboard the Shenzhou 5 spacecraft (the word *Shenzhou* translating to 'divine vessel') on 15 October 2003. The carrier rocket was a Long March 2F, and the launch took place from the Jiuquan Satellite Launch Center, based in the Gobi Desert in Inner Mongolia.

Having fulfilled that first crucial step, the second phase of the Chinese space programme involved the development of advanced space flight techniques and technologies, which would include EVAs by the taikonauts as well as orbital rendezvous and docking. One critical aspect of this second phase of Project 921 was the development of prototype Tiangong (Heavenly Vessel) orbiting space stations. Tiangong-1 was

Yang Liwei became
the first Chinese
national to fly
into space aboard
Shenzhou 5 in
October 2003.

launched in September 2011 and used by crews to practise rendezvous and docking manoeuvres during short-term space missions. These docking manoeuvres were carried out by the crews of Shenzhou 7, 9 and 10 (Shenzhou 8 was an unmanned, automatic docking flight).

By March 2016, ground controllers had lost all communication with Tiangong-1, but in its orbital lifetime the station provided much-needed design and systems information that would be incorporated in the follow-up space station, Tiangong-2, launched in September of that year. It featured a far greater capacity for scientific experimentation, in addition to such comforts as living quarters, cargo transport, refuelling and replenishment, and infrastructure that would allow for lengthy human habitation. In this way, the two-man crew of Shenzhou 11 were able to live aboard Tiangong-2 for thirty days in late 2016.[13] The vacated, dormant Tiangong-1 would continue orbiting Earth until 2 April 2018, when it was commanded to re-enter the atmosphere. Those segments that failed to burn up in the heat of re-entry splashed down across the South Pacific. In turn, Tiangong-2 was similarly de-orbited in July 2019. A third Tiangong space station, which would have supported extended

## Crewed Chinese Space Missions To Date

| Taikonaut(s) | Mission | Launch | Landing | Mission Highlights |
|---|---|---|---|---|
| Yang Liwei | Shenzhou 5 | 15 October 2003 | 15 October 2003 | First Chinese national in space |
| Fei Junlong Nie Haisheng | Shenzhou 6 | 12 October 2005 | 16 October 2005 | First Chinese two-man space flight |
| Jing Haipeng Liu Boming Zhai Zhigang | Shenzhou 7 | 25 September 2008 | 28 September 2008 | First three-man Chinese space flight. First EVA by a Chinese national (Zhai) |
| Uncrewed | Shenzhou 8 | 31 October 2011 | 17 November 2011 | Automatic docking with Tiangong-1 space station |
| Jing Haipeng Liu Wang Liu Yang | Shenzhou 9 | 16 June 2012 | 29 June 2012 | First Chinese crew to achieve manned spacecraft docking with Tiangong-1. First repeat taikonaut (Jing). First Chinese woman in space (Liu Yang) |
| Nie Haisheng Zhang Xiaoguang Wang Yaping | Shenzhou 10 | 11 June 2013 | 26 June 2013 | Second manned space rendezvous and docking with Tiangong-1. Second Chinese woman in space (Wang) |
| Jing Haipeng Chen Dong | Shenzhou 11 | 17 October 2016 | 18 November 2016 | Docked with Tiangong-2 space station. Third flight for mission commander Jing |

occupation by several crews, was already on the drawing board, but the programme was eventually cancelled.

To date, China has launched eleven taikonauts into space aboard six manned Shenzhou spacecraft, as well as one cargo spacecraft, and space stations Tiangong-1 and Tiangong-2, thus completing the first two steps of the country's manned space programme. For China, the next, third step will be to assemble and operate a permanently manned modular space station. Although details are scarce and vague, this station is expected to be completed around 2022 and will support their long-term goals for space exploration, including missions to the Moon and Mars. China is now in the process of selecting a new cadre

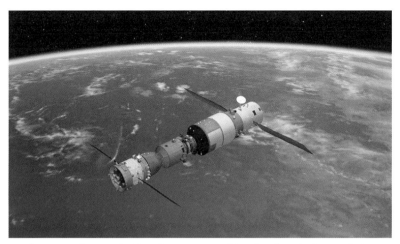

An illustration showing the Shenzhou 11 spacecraft heading for a docking with the Tiangong-2 space station.

On the Shenzhou 9 mission in June 2012, Liu Yang became the first Chinese woman in space.

of taikonaut trainees for future space station missions, as well as developing a new-generation, much larger spacecraft. China has plans to launch its new-generation spacecraft on an unmanned test flight sometime before 2022, flying into orbit atop a Long March 5B heavy-lift rocket. China realizes that to get the spacecraft the rest of the way to its lunar destination, it first needs to develop a more powerful rocket

– the super-heavy-lift Long March 9. For this reason, the country is targeting a crewed mission to the Moon only sometime 'in the 2030s' (although some Chinese sources have suggested earlier dates).[14]

## The Challenge of Space 'Tourism'

In his 1958 book *Have Space Suit – Will Travel*, renowned science fiction author Robert Heinlein stated: 'The idea that any rich man could simply lay cash on the line and go [into space] was more than I could stand.' Yet in April 2001 a slight, grey-haired sixty-year-old engineer did just that. Dennis Tito, who had amassed a fortune through his investment management consultancy firm, paid around U.S.$20 million to the Russian Air and Space Agency, Rosaviakosmos (also known as Roscosmos), through Virginia-based Space Adventures, Inc., and quickly garnered celebrity status as the world's first private space 'tourist'.

As a teenager in 1957, Dennis Tito had been inspired by the launch of the first Sputnik satellite. He subsequently earned degrees in aeronautical and aerospace engineering, and then spent five years working at NASA's Jet Propulsion Laboratory (JPL) in California, helping to design trajectories for the Mariner Mars and Venus probes. In 1972 he entered the world of finance and began using his mathematical and engineering skills to develop new approaches in his analysis of stock market risks. By 1998, his Wilshire Associates was the third-largest investment management consultancy firm in the United States.

Tito was originally scheduled to fly to Russia's Mir space station, but once the difficult decision had been made to de-orbit Mir, plans were revised to send him instead to the semi-constructed International Space Station. It was an intrusive concept that NASA did not like at all. In fact, there was a face-off between NASA and Russia regarding Tito's self-funded adventure. NASA argued that Tito was not fully trained and would therefore present a safety risk. The ISS crew, they said, would have to perform distracting 'babysitting' duties to ensure he didn't do anything that might endanger their lives or harm the station. Additionally, with the modular ISS still undergoing assembly, it was a highly inappropriate time to bring aboard some unwelcome, camera-wielding tourist.

The Russians argued that, under the provisions of their intergovernmental agreement (IGA), they had every right to select any crew members they wanted, as long as they were fully trained. More to the point, of course, they were in desperate need of Tito's millions.

Dennis Tito (left) with Russian crewmates Talgat Musabayev and Yuri Baturin aboard the ISS in 2001.

In the full knowledge that the welcome mat would not be laid out for him when his Soyuz TM-32 spacecraft docked with the ISS, Tito nevertheless continued with his training.[15] Faced with the inevitability of the situation, NASA imposed a strict condition on the Russian space agency, requiring them to compensate other space station partners for any damage to their equipment resulting from Tito's activities. Fortunately, the agreement was never tested, as there were no reports of damage during Tito's seven-day visit to the ISS. He had been launched aboard Soyuz TM-32 on 28 April 2001 and would spend seven of his ten days of space flight aboard the ISS. Despite overt criticism at the time, what his flight did succeed in doing was to dramatically increase public awareness and enthusiasm for so-called 'space tourism'.

In the following years, several other paying 'space flight participants' (as they preferred to be known) would also fly aboard a Soyuz spacecraft to the orbiting ISS. Each would have undergone several months of physical and language training in Russia for their flight. With only one seat available on selected Soyuz flights, a prime participant candidate was selected along with a backup, who would take over the seat in the event of any last-minute problems or illness.

Although in most cases the prime candidate flew into space, others were unlucky to miss out. In February 2002, 22-year-old Lance Bass from the boy band NSYNC announced that he had agreed to pay approximately U.S.$20 million for a November flight to the ISS on Soyuz TMA-1.

His only previous qualification for flying into space was that he had attended Space Camp in Alabama at the age of twelve. He then began his training at Moscow's Star City cosmonaut training centre and later at the Johnson Space Center in Texas. In August, Bass's group of financial backers, which included MTV and RadioShack, missed the payment deadline for the money. The following month, some very displeased Russian space officials told Bass that ongoing payment discussions had stalled yet again. Conveniently, they also revealed that Bass had been found to be suffering from 'a minor heart ailment', and as a consequence he was bumped from the flight. At just 23, he would have been the youngest person ever to have flown into space.[16]

Thirteen years later, another well-known singer, Sarah Brightman of *Phantom of the Opera* fame, began training for a self-funded ten-day Soyuz mission to the ISS and back, scheduled for 1 September 2015. The U.S.-based space tourism company Space Adventures was brokering the multimillion-dollar flight for the British singer. To the surprise of many, on 13 May Brightman announced that she would not be making that trip to orbit. An update on her website stated: 'Ms Brightman said that, for personal family reasons, her intentions have had to change, and she is postponing her cosmonaut training and flight plans at this time. She

## Self-funded Space Flight Participants

| Participant | Backup | Flight to/from ISS | Launch | Return |
|---|---|---|---|---|
| Dennis Tito, 60, USA | — | Soyuz TM-32/ Soyuz TM-31 | 28 April 2001 | 6 May 2001 |
| Mark Shuttleworth, 28, South Africa | — | Soyuz TM-34/ Soyuz TM-33 | 25 April 2002 | 5 May 2002 |
| Gregory Olsen, 60, USA | — | Soyuz TMA-7/ Soyuz TMA-6 | 1 October 2005 | 11 October 2005 |
| Anousheh Ansari, 40, Iran/USA | — | Soyuz TMA-9/ Soyuz TMA-8 | 18 September 2006 | 29 September 2006 |
| Charles Simonyi, 58, Hungary/USA | — | Soyuz TMA-10/ Soyuz TMA-9 | 7 April 2007 | 21 April 2007 |
| Richard Garriott, 47, USA | Nik Halik, 39, Australia | Soyuz TMA-13/ Soyuz TMA-12 | 12 October 2008 | 23 October 2008 |
| Charles Simonyi, 60, Hungary/USA | Esther Dyson, 47, Switzerland/USA | Soyuz TMA-14/ Soyuz TMA-13 | 26 March 2009 | 8 April 2009 |
| Guy Laliberté, 49, Canada | Barbara Barrett, 58, USA | Soyuz TMA-16/ Soyuz TMA-14 | 30 September 2009 | 11 October 2009 |

Although not self-funded, other space flight participants enjoyed sponsored visits to the ISS, with funding provided by agreement with their home nation.

## Nationally Funded Space Flight Participants

| Name | Backup | Flight to/from Mir | Launch | Return |
|---|---|---|---|---|
| Marcos Pontes, Brazil | — | Soyuz TMA-8/ Soyuz TMA-7 | 30 March 2006 | 8 April 2006 |
| Sheikh Muszaphar Shukor, Malaysia | Faiz Khaleed | Soyuz TMA-11/ Soyuz TMA-10 | 10 October 2007 | 21 October 2007 |
| Yi So-yeon, South Korea | Ko San | Soyuz TMA-12/ Soyuz TMA-11 | 8 April 2008 | 19 April 2008 |

would like to express her extreme gratitude to Roscosmos, Energia, GCTC [Gagarin Cosmonaut Training Center], Star City, NASA and all the cosmonauts and astronauts for their support during this exciting time in her life.' Brightman's now vacant third seat on the Soyuz TMA-18M flight to the ISS was subsequently handed to cosmonaut Aidyn Aimbetov from KazCosmos, the Kazakh space agency.[17]

The history of human space exploration will continue unabated. On 25 September 2019, the first Emirati astronaut, Hazza Al Mansouri, journeyed to the International Space Station, aboard the Soyuz MS-15 spacecraft. His crew consisted of Russian commander Oleg Skripochka (making his third space flight) and NASA astronaut and flight engineer Jessica Meir. Al Mansouri would become the first Arab space traveller to occupy the ISS. Eight days later, on 3 October, he landed safely in Kazakhstan along with the crew of Soyuz MS-12. Denmark also has plans for a similar manned space flight, possibly in 2022.

In another developing story, the Indian Space Research Organisation (ISRO) has been working since 2007 on developing the technology necessary to launch crewed spacecraft into orbit. The first such flight, sending a crew of two or three *gagan yatri* (Sanskrit for 'sky travellers') on a seven-day mission, was planned for December 2021, but will likely occur sometime in 2022. They will be aboard a spacecraft called *Gaganyaan* (Sky Vehicle) and will be launched aboard a GSLV Mark III rocket. If these plans come to fruition, India will become the fourth nation to conduct independent human space flight after the Soviet Union/Russia, the United States and China. After conducting crewed space flights, ISRO further intends to develop India's own space station programme, leading eventually to a possible crewed lunar landing.[18]

In his 2018 Independence Day address, India's prime minister Narendra Modi formally announced that a manned *Gaganyaan* mission would take place by 2022, to celebrate the nation's 75th year of independence. By November 2019 the Indian Air Force had selected twelve potential *gagan yatri* to travel to Russia for training. The following month, the list of candidates had been narrowed to four, and in February 2020 they began far more intensive training. Athough the names of three of the candidates had not been announced at the time of writing, it was revealed in August 2020 that Wing Commander Nikhil Rath of the Indian Air Force had completed his training for the flight.[19]

# EPILOGUE

# OUR FUTURE IN SPACE

O n 21 June 2004, 62-year-old civilian test pilot Mike Melvill entered the history books as the first person to pilot a private craft in space. The vehicle he flew – SpaceShipOne (ss1), built by cutting-edge aeronautical engineer Burt Rutan's company Scaled Composites – also made history as the first private manned spacecraft.

The flight was the first of two needed to win a $10 million incentive offered by the X Prize Foundation, a non-profit organization founded under the leadership of Greek American billionaire and engineer, physician and entrepreneur Peter Diamandis. The foundation had been created in May 1996 with the aim of inspiring innovation through competition and to promote the development of a low-cost, efficient craft for space tourism, in much the same way as companies offering cash prizes were used to stimulate commercial aviation in the early twentieth century. Initially known as the X Prize, the race for private space flight was renamed the Ansari X Prize on 5 May 2004, the 43rd anniversary of Alan Shepard's suborbital space flight of 1961, following a multimillion-dollar donation from Iranian-born entrepreneurs Anousheh Ansari and Amir Ansari. (Two years later, in September 2006, Anousheh Ansari would fly to the ISS as a paying space tourist.) In order to win the Ansari X Prize, a competing private team (26 teams had registered) had to exceed 100 km (62 mi.) altitude twice in a two-week period using the same spacecraft, with one pilot and two passengers (or equivalent ballast) on board.[1] Unfortunately, control problems experienced during the flight caused the second flight to be abandoned, and a new attempt would be delayed until 29 September.

Virgin Galactic's Sir Richard Branson.

Paul Allen, Microsoft co-founder-turned-investor and philanthropist, had decided to bankroll the Scaled Composites project, joining forces with Burt Rutan. The legendary designer and aeronautical genius came up with SpaceShipOne, an experimental air-launched suborbital spaceplane that would be carried aloft by its mothership, the futuristic twin-turbojet White Knight.

On 27 September, as SpaceShipOne was being readied for its second attempt at the X Prize, the chairman of Virgin Atlantic Airways, Sir Richard Branson, announced a contract with Rutan and Microsoft financier Allen that gave Scaled Composites funding to design and build a five- to eight-passenger vehicle, and a licence to use the technology for commercial suborbital flights.

The first attempt – the sixteenth flight of SpaceShipOne – took place on 29 September 2004, with pilot Mike Melvill flying to a height of 102.93 km (63.96 mi.), thereby exceeding the required altitude. A week later, Branson was at hand in the Mojave Desert to witness the flight that it was hoped would secure the Ansari X Prize and the $10 million. Flight 17 began on 4 October – coincidentally the 47th anniversary of the launch of the first Sputnik satellite – and this time pilot Brian Binnie was aboard SpaceShipOne as it was carried aloft by the White Knight. The craft was dropped by the mothership, the engines fired, and the spacecraft soared to a record altitude of 112.014 km (69.6 mi.).

Peter Diamandis could barely contain his excitement: 'Today we make history,' he told assembled reporters. 'Today we go to the stars!'[2]

Within a month of announcing his plans for future commercial suborbital space flights, Branson's new offshoot company Virgin Galactic had already received some 7,000 strong expressions of interest, a number that continued to soar. Asked to comment on his grand scheme, he said: 'We hope to create thousands of astronauts over the next few years and bring alive their dream of seeing the majestic beauty of our planet from above, the stars in all their glory and the amazing sensation of weightlessness. The development will also allow every country in the world to have their own astronauts, rather than the privileged few.'[3]

Initially, Branson suggested that the first passenger-carrying space flight might occur as soon as 2009, but this projected date continued to slide. In 2009 Virgin Galactic continued to be upbeat, predicting that initial flights would take place from the Mojave Desert's Spaceport America 'within two years'. On 7 December that year a second craft, SpaceShipTwo, was unveiled at the Spaceport, with Branson telling the assembled crowd of three hundred people, many of whom had paid $200,000 apiece to book passage on one of his flights, that he expected the first flights to take place in 2011.

A triumphant Brian Binnie stands on top of SpaceShipOne after winning the Ansari X Prize on 4 October 2004.

Delays continued to plague the project, the most serious of which was the loss of SpaceShipTwo, named vss *Enterprise* (Virgin Space Ship *Enterprise*), on 31 October 2014, during the craft's fourth rocket-powered test flight. Shortly after being released from the mothership, it broke apart in mid-air. The co-pilot, Michael Alsbury, was killed in the subsequent crash, while the pilot, Peter Siebold, was seriously injured. The later investigation found that the air braking system appeared to have been deployed too early and incorrectly by Alsbury, causing SpaceShipTwo to break up.[4]

A replacement spaceship, vss *Unity*, completed its first test flight in December 2016, and two years later achieved an altitude of 82.7 km (51.4 mi.) above Earth, reaching close to three times the speed of sound. The two pilots on this flight were Mark Stucky and former NASA astronaut Frederick Sturckow, who were presented with commercial astronaut wings by the U.S. government for their accomplishment.

With vss *Unity* undergoing interior refurbishment and other upgrades, Virgin Galactic continued to build on its fleet of spaceships, hoping to have constructed a fleet of five vehicles by 2023. In 2020 there were plans for sixteen suborbital flights to take place that year, but in accordance with guidelines issued by the New Mexico Department of Health in response to the spread of the COVID-19 virus, Virgin Galactic

vss *Unity* (previously referred to as vss *Voyager*) is the second suborbital SpaceShipTwo to be built.

have minimized their operations and a new flight timetable will be issued once it is considered safe to do so.

Virgin Galactic is not the only player in town, with Elon Musk's SpaceX and Jeff Bezos's Blue Origin among the many entrepreneurial companies looking to a future involving private space flight. Early on, with all this burgeoning interest raising public awareness, it was time for concerned lawmakers to step in. In December 2004, a Commercial Space Launch Amendments Act was signed into law in the United States by 269 votes to 120. Aimed at promoting the development of the emerging commercial human space flight industry, it also included provisions that would allow paying passengers to fly at their own risk. Two months later, the Washington-based Federal Aviation Administration (FAA) reacted by publishing draft guidelines aimed at granting permits to those companies wishing to test reusable suborbital rockets. These guidelines were also aimed at governing crews and passengers aboard private spacecraft. The FAA's authority to govern the industry with binding regulations was seen as a significant step forward, allowing for government oversight while permitting the industry to move towards a self-policing model.[5]

As a direct result, a dozen companies interested in private rocket flights established the Personal Spaceflight Federation (now the Commercial Spaceflight Federation) in February 2005, which would work with federal regulators to help advance the infant industry's interests. The founders included Diamandis, Rutan and Eric Anderson, co-founder and chairman of Space Adventures. Also on board was video game innovator John Carmack, whose Armadillo Aerospace team was at one time a leading contender for the X Prize.

Entrepreneur Gregg Maryniak was the co-founder of the X Prize Foundation and its original executive director, and he served as the Personal Spaceflight Federation's chief spokesman. At the time he told MSNBC.com that the new legislation 'basically says that government and industry need to work together to come up with safety standards, and we believe very strongly that industry's going to have to come up with a product that's considerably safer than the current civil space program.' He also cited several surveys indicating that 70 per cent of Americans would buy a flight into space 'if they had the chance and the price was right'.[6] Greg Autry, a space flight enthusiast and lecturer at the University of California, offered his opinion that tourists undertaking those brief rocket rides to the very rim of space may actually be bewildered by the speed of their experience. 'Although this will clearly attract a lot of "extreme" folks, they will likely be surprised by the intensity of this ride,' he commented. 'There is virtually no time to gather your wits and stomach to enjoy the view before you go right back down.'[7]

### To See Earth from Space

As this book heads towards publication, plans for future space exploration are being generated and tested. In addition to NASA's plans to return astronauts to the Moon and beyond, a number of well-heeled entrepreneurs are lining up and spending their considerable wealth in creating vehicles and rockets that will carry people into space.

When Neil Armstrong and Buzz Aldrin walked on the Moon in 1969, SpaceX CEO Elon Musk had not even been born; Blue Origin owner Jeff Bezos was just five years old; and Virgin Galactic's Sir Richard Branson had turned nineteen just two days earlier. Among many others, these are the principal visionaries in the thrust to open space to extended exploration and even commercial tourism.

The Spaceship Company (TSC) is Virgin Galactic's space-system manufacturing organization, headquartered at the Mojave Air and Space Port in California. They are currently building and testing a fleet of White Knight Two carrier aircraft and SpaceShipTwo reusable spaceships, which together form Virgin Galactic's human space flight system. Although it has been many years in the realization of Branson's dream, more than six hundred people from around sixty countries have already signed up to pay the equivalent of around £200,000 in order to take a brief journey into space and experience a couple of minutes of weightlessness as they gaze out through the windows of a SpaceShipTwo vehicle at the curvature of Earth below them. Or, as Branson has pledged to his customers, 'a unique, multi-day experience, culminating in a personal spaceflight that includes out-of-seat zero gravity and views of Planet Earth from space'.[8]

The first of these customer space flights has been delayed over and over again as the years have gone by, especially following the fatal crash involving SpaceShipTwo and then strict health limitations imposed by the spread of the COVID-19 virus.

In 2002 South African-born billionaire Elon Musk founded Space Exploration Technologies – better known today as SpaceX. Musk was a co-founder in 1999 of the X.com company, which later evolved into

The Scaled Composites SpaceShipTwo spaceplane (central fuselage) resting under its mother ship, White Knight Two, inside a hangar in Mojave, California.

the electronic payment firm PayPal, and is also the co-founder and CEO of electric car manufacturer Tesla. Passionate about space exploration, he has overseen the development and manufacture of highly advanced but price-competitive rockets and spacecraft suitable for missions into orbit and further out into the cosmos. The maiden flight of his first rocket, Falcon 1, took place in 2006, and of the larger Falcon 9 in 2010.

In February 2018, Musk's third rocket, Falcon Heavy, was launched for the first time. It was designed to carry a load of up to 53,000 kg (117,000 lb) into orbit. On this test launch, the Falcon Heavy carried an unusual payload – a cherry-red Tesla Roadster, not only equipped with cameras to provide some epic views for the vehicle's planned orbit around the Sun but carrying a space suit-clad dummy 'driver' named Starman, after the David Bowie song.

SpaceX has also been working on the development of the Dragon spacecraft, designed to carry supplies to the ISS as well as up to seven astronauts. The company is also developing fully reusable rockets with the ability to lift off and return vertically to the launch pad. In July 2018, a new Block 5 Falcon rocket successfully touched down on a drone ship less than nine minutes after launch.

Musk also has an eye on space tourism, and in 2018 SpaceX announced that a Japanese billionaire, Yusaku Maezawa, had already booked a seat on a trip around the Moon for himself and a couple of

Elon Musk's Tesla Roadster, with Earth in the background. The 'Starman' mannequin wearing a SpaceX spacesuit is seated in the driving seat. Camera is mounted on an external boom.

An artist's rendition of a SpaceX Crew Dragon docking with the ISS as it will during a mission for NASA's Commercial Crew Program.

other guests, aboard a SpaceX Starship rocket (formerly known as the Big Falcon Rocket, or BFR), which at this moment in time is still under development and testing.

In January 2020 an unmanned Dragon spacecraft completed an in-flight abort test, disengaging from its Falcon 9 rocket less than 90 seconds after launch from the Kennedy Space Center, splashing down under parachutes into the ocean around 32 km (20 mi.) off the coast

A SpaceX Falcon Heavy rocket is ready for launch on the pad at Launch Complex 39A at the Kennedy Space Center, Florida, on 24 June 2019.

of Florida. It was a crucial step on the way to the first planned crewed mission to the ISS. The next step, designated Crew Dragon Demo-1, began with the launch of an uncrewed Dragon spacecraft on 2 March 2020, which successfully docked with the ISS the following day. Five days later the empty spacecraft was undocked from the ISS and following re-entry was recovered intact from the Atlantic Ocean.

The following mission, Demo-2, was launched on 30 May 2020 and this time carried a crew of two NASA astronauts – Bob Behnken and Doug Hurley – on a test flight to and from the ISS. It marked the first time American astronauts had been launched into orbit aboard a commercially built and operated American-crewed spacecraft since 2011. Following a successful docking with the ISS, Behnken and Hurley spent two months aboard the orbiting station before reboarding their Dragon spacecraft (which they had named *Endeavour*) and completing a successful re-entry and splashdown on 2 August.

Another historic flight took place with a successful night-time launch of the Crew-1 mission from the Kennedy Space Center on 15 November

Elon Musk explains the capabilities of Starship to NORAD and U.S. Air Force Space Command in April 2019.

2020. The Falcon 9 launch vehicle and Dragon spacecraft, named *Resilience* by the NASA crew of Michael Hopkins, Victor Glover, Shannon Walker and JAXA astronaut Soichi Noguchi, performed flawlessly. *Resilience* docked with the ISS the following day after a 27-hour orbital chase, becoming the 100th crewed vehicle to arrive at the orbital complex. Once the hatches had been opened they were greeted by the resident crew of cosmonauts Sergei Ryzhikov (station commander) and Sergei Kud-Sverchkov and NASA astronaut flight engineer Kate Rubins. The Dragon crew would remain on board the station for six months before returning to a splashdown, following which the spacecraft could be refurbished for another mission.

Another company looking to space is aerospace manufacturer Blue Origin, founded and privately owned by entrepreneur Jeff Bezos, said to be the richest man in the world as the founder and CEO of Amazon. com and owner since 2013 of the *Washington Post* newspaper. Blue Origin is developing its *New Shepard* rocket specifically for short trips

This illustration made available by the SpaceX company depicts the Crew Dragon capsule and Falcon 9 rocket during a launch from the Kennedy Space Center in Florida.

The crew of SpaceX Dragon spacecraft *Endeavour*'s test flight to the ISS: NASA astronauts Bob Behnken (left) and Doug Hurley. They were on the first crewed mission to be launched into orbit from the United States since 2011.

into space carrying paid tourists, as well as the heavy-lift *New Glenn* rocket for satellite launch contracts and space tourism purposes. Both rockets were named after pioneering Mercury astronauts. As with Virgin Galactic and SpaceX, New Origin hopes to be launching people into space once the scourge of the COVID-19 virus is behind us and the company can set a new launch timetable. Bezos has also announced

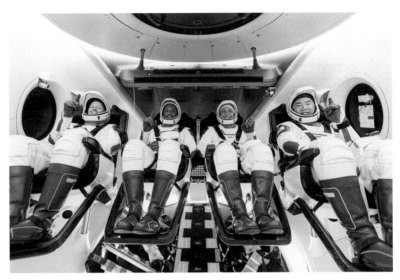

NASA's SpaceX Crew-1 crew members seated in the company's spacious Crew Dragon spacecraft during training. Left to right: NASA astronauts Shannon Walker, Victor Glover and Mike Hopkins, and JAXA astronaut Soichi Noguchi.

The successful launch of one of Blue Origin's *New Shepard* rockets.

Jeff Bezos, founder and owner of Blue Origin.

plans for the development of what he calls the Blue Moon lunar lander, capable of carrying space travellers, cargo and supplies to the lunar surface, possibly as early as 2023.

NASA, of course, is still a major player in human space exploration and is committed to landing the next generation of astronauts on the Moon. It has already been announced by the space agency and subsequently reinforced by former president Donald Trump that the next person to set foot on the Moon will be a woman. This new programme will fly under the banner of Artemis, named after the twin sister of Apollo and the goddess of the Moon in Greek mythology. As plans currently stand, NASA will launch astronauts aboard an Orion spacecraft atop their powerful new rocket, known as the Space Launch System, or STS. On arrival in lunar orbit, the astronauts will dock Orion to the previously positioned Gateway lunar satellite before later proceeding down to the lunar surface. As the space agency has explained:

> NASA is working with its partners to design and develop a small spaceship that will orbit the Moon called the Gateway. This spaceship will be a temporary home and office for astronauts [and] will have living quarters, laboratories for science and research, docking ports . . . for visiting spacecraft, and more. It will provide NASA and its partners access to more of the lunar surface than ever before, supporting both human and robotic missions. The Gateway will be our home base for astronaut expeditions on the Moon, and future human missions

Lockheed Martin has completed the construction of the Orion spacecraft that will circumnavigate the Moon as part of the Artemis programme.

to Mars. Even before our first trip to Mars, astronauts will use the Gateway to train for life far away from Earth, and we will use it to practice moving a spaceship in different orbits in deep space . . . Astronauts will visit the Gateway at least once per year, but they won't stay year-round like crew aboard the International Space Station . . . Its interior is about the size of a studio apartment (whereas the space station is larger than a six-bedroom house). Once docked, astronauts can live and work aboard the spaceship for up to three months at a time, conduct science experiments, and take trips to the surface of the Moon.[9]

NASA is currently working towards the launch of Artemis 1, an uncrewed test flight of the SLS and the Orion spacecraft. This will be followed by Artemis 2, planned as the first crewed test mission. The space agency's ambitious target for this flight is currently no earlier than August 2023, with further plans to launch the first crew to the Moon in late 2024 on the Artemis 3 mission. Further flights are then planned approximately once each year. Flights to Mars, although well in the future, are still on the drawing board.

As well as NASA's plans, we are without doubt on the verge of an exciting phase of space exploration – our Greatest Adventure. As

Christian Davenport so succinctly wrote in his book *The Space Barons* (2018), which looked at the past and future accomplishments of space flight visionaries such as Elon Musk and Jeff Bezos: 'Their race to the stars was driven not by war or politics, rather by money and ego and adventure, a chance to extend humanity out into space for good.'[10]

We are poised to see – and for some to participate in – the next phase of our delayed human journey out into the perilous beauty of the cosmos. It will not be easy. To quote another source, the movie *All about Eve* (1950): 'Fasten your seatbelts, it's going to be a bumpy ride.'

# REFERENCES

## PROLOGUE

1 Robert Forsyth, *Bachem Ba 349 Natter* (London, 2018), p. 7.
2 Wernher von Braun, excerpt from speech given at the Sixteenth National Conference on the Management of Research, French Lick, Indiana, 18 September 1962. From 'Retrospective: A Speech by Wernher von Braun on Management', available at https://medium.com, 22 September 2015.
3 John F. Kennedy, Department of State Central Files, National Archives, Washington, DC, 711.11-KE/2-1261. Also printed in *Public Papers of the Presidents of the United States: John F. Kennedy, 1961*, p. 257, available at www.quod.lib.umich.edu.
4 John F. Kennedy, news conference held at State Department Auditorium, Washington, DC, 12 April 1961, available at www.jfklibrary.org.
5 John F. Kennedy, excerpt from *Address before a Joint Session of Congress*, 25 May 1961, available at www.jfklibrary.org.

## 1 THE REALITY OF TOMORROW

1 James Harford, *Korolev: How One Man Masterminded the Soviet Drive to Beat America to the Moon* (New York, 1997).
2 Colin Burgess and Chris Dubbs, *Animals in Space: From Research Rockets to the Space Shuttle* (Chichester, 2007), pp. 25–6.
3 Rex Hall and David J. Shayler, *The Rocket Men: Vostok and Voskhod, the First Soviet Manned Spaceflights* (Chichester, 2001), p. 25.
4 Burgess and Dubbs, *Animals in Space*, pp. 64–8.
5 A. V. Pokrovsky, 'Vital Activity of Animals during Rocket Flights into the Upper Atmosphere', in *Behind the Sputniks: A Survey of Soviet Space Science*, ed. F. J. Krieger (Washington, DC, 1958). Originally presented as a report to the International Congress on Guided Missiles and Rockets, Paris, 3 8 December 1956.
6 Burgess and Dubbs, *Animals in Space*, p. 83.
7 Sir Bernard Lovell quoted in Paul Dickson, *Sputnik: The Shock of the Century* (New York, 2001), p. 24.
8 Robert Silverberg, *First American into Space* (Derby, CT, 1961), p. 31.
9 Science Correspondent, 'Five Years of the Space Age: Benefits and Dangers – IGY Aims Submerged by Cold War', *The Guardian*, 5 October 1962, p. 10.
10 Dwight D. Eisenhower quoted in 'Impact of Russian Satellite to Boost U.S. Research Effort', *Aviation Week*, XIV (October 1957), pp. 28–9.
11 Anatoly Zak, 'Sputnik-2 in Orbit', www.russianspaceweb.com, updated 2 November 2017.

12 Colin Burgess oral interviews with Dr Oleg Gazenko and Vitaly Sevastyanov, ASE (IX) Congress, Vienna, Austria, 10–17 October 1993.

13 Oleg Ivanovskiy (writing as Alexei Ivanov), *Vpervyye: zapiski vedushchego konstruktova* [in Russian; The First Time: Notes of a Leading Designer], 2nd edn (Moscow, 1982).

14 Colin Burgess and Simon Vaughan, 'America's First Astro-chimps', *Spaceflight* (British Interplanetary Society), XXXVIII (July 1996), pp. 236–8.

15 Helen C. Allison, 'News Roundup', *Bulletin of the Atomic Scientists: A Magazine of Science and Public Affairs*, XIV/3 (March 1958), p. 126.

16 Maj. Gen. John B. Medaris, U.S. Army: Testimony Given at Inquiry into Satellites and Missile Programs, Hearings before the United States Senate Committee on Armed Services, 85th Congress, Washington, DC, 14 December 1957. From *History of Acquisition in the Department of Defense*, vol. I: *Rearming for the Cold War 1945–1960*, Elliott V. Converse III for the Historical Office of the Secretary of Defense, Washington, DC (2012), p. 572.

17 Garrison Norton, U.S. Asst. Secretary of the Navy for Air: Testimony Given at Inquiry into Satellites and Missile Programs, Hearings before the United States Senate Committee on Armed Services, 85th Congress, Washington, DC, 16 December 1957 (First and Second Sessions, Part 2, Page 721). Published by U.S. Government Printing Office (Washington, DC), 1958.

18 Kurt Stehling quoted in Dickson, *Sputnik*, p. 156.

19 Helen T. Wells, Susan H. Whiteley and Carrie E. Karegeannes, *Origins of NASA Names* (Washington, DC, 1976), p. 106.

20 Burgess and Dubbs, *Animals in Space*, pp. 137–40.

21 'Soviet Space Medicine, Smithsonian Videohistory Program, with Abraham Genin', Smithsonian Videohistory Program. Cathleen S. Lewis interviewer, 29 November 1989, available at Smithsonian Institution Video Archives, Record Unit 9551.

## 2 FIRST INTO OUTER SPACE

1 Colin Burgess, *Selecting the Mercury Seven: The Search for America's First Astronauts* (Lincoln, NE, 2011), pp. 274–8.

2 Anon., 'USS Donner LSD 20 Recovery Ship MR2 with Space Chimpanzee Ham', *Gator News* (Amphibious Force, U.S. Atlantic Fleet, Little Creek, VA), XIX, 3 February 1961.

3 Yaroslav Golovanov, *Our Gagarin: The First Cosmonaut and His Native Land*, trans. David Sinclair-Loutit (Moscow, 1978), pp. 44–55.

4 *New York Times*, 15 March 1961, p. 8, citing Sergei Khrushchev: *Nikita Khrushchev: krizisy i rakety: uzglyad iznutri* [Crises and Missiles: An Inside Look], vol. 1 (Moscow, 1994), p. 97.

5 Robin McKie, 'Sergei Korolev: The Rocket Genius behind Yuri Gagarin', *The Guardian*, March 2011, p. 11.

6 Joint Publication Research Service, 'Manned Mission Highlights', in *Report on Science and Technology – Central Eurasia: Space* (JPRS-USP-92-004) (Springfield, VA, 10 June 1992).

7  Francis French and Colin Burgess, *Into That Silent Sea: Trailblazers of the Space Era, 1961–1965* (Lincoln, NE, 2007), p. 54.

8  James Schefter, *The Race: The Uncensored Story of How America Beat Russia to the Moon* (New York, 1999), p. 112.

9  Dee O'Hara, telephone interview with the author, 22 May 2002.

10  French and Burgess, *Into That Silent Sea*, p. 59.

11  M. Scott Carpenter, Gordon L. Cooper, John H. Glenn, Virgil I. Grissom, Walter M. Schirra, Alan B. Shepard and Donald K. Slayton, *We Seven: By the Astronauts Themselves* (New York, 1962), p. 241.

12  French and Burgess, *Into That Silent Sea*, p. 62.

13  Howard Benedict, Jay Barbree, Alan Shepard and Deke Slayton, *Moon Shot: The Inside Story of America's Apollo Moon Landings* (Nashville, TN, 1994), p. 78.

14  'U.S. Hurls Man 115 Miles into Space: Shepard Works Controls in Capsule, Reports by Radio in 15-Minute Flight', *New York Times*, 6 May 1961, p. 1.

15  Telegram from USSR Chairman Nikita Khrushchev to U.S. President John Kennedy, 6 May 1961. Department of State Central Files, National Archives, Washington, DC, 911.802/5-661.

16  John F. Kennedy, *Excerpt from Address before a Joint Session of Congress*, 25 May 1961, John F. Kennedy Presidential Library and Museum, Boston, MA.

17  Curt Newport, *Lost Spacecraft: The Search for Liberty Bell 7* (Burlington, ON, 2002), pp. 164–73.

18  Michelle Evans: *The X-15 Rocket Plane: Flying the First Wings into Space* (Lincoln, NE, 2013).

19  Frederick A. Johnsen, 'X-15 Pioneers Honored as Astronauts', 23 August 2005, www.nasa.gov/missions/research/index.html.

**3 INTO ORBIT**

1  Loyd S. Swenson Jr, James M. Grimwood and Charles C. Alexander, 'MR-1: The Four-inch Flight', in *This New Ocean: A History of Project Mercury*, NASA History Series SP-4201 (Washington, DC, 1989), available at https://history.nasa.gov.

2  Rex Hall and David J. Shayler, *The Rocket Men: Vostok and Voskhod, the First Soviet Manned Spaceflights* (Chichester, 2001), pp. 174–5.

3  Lester A. Sobel, ed., *Space: From Sputnik to Gemini* (New York, 1965), p. 126.

4  Colin Burgess, *Friendship 7: The Epic Orbital Flight of John H. Glenn, Jr* (Chichester, 2015), p. 58

5  Colin Burgess, *Selecting the Mercury Seven: The Search for America's First Astronauts* (Chichester, 2011), pp. 279–80.

6  Burgess, *Friendship 7*, p. 23.

7  Colin Burgess and Chris Dubbs, *Animal Astronauts: From Research Rockets to the Space Shuttle* (Chichester, 2007), pp. 264–8.

8  Sobel, ed., *Space*, p. 160.

9  Anon., 'John Glenn: One Machine That Worked without Flaw', *Newsweek*, 5 March 1962, p. 24.

10  Burgess, *Friendship 7*, p. 149.

11 Colin Burgess, *Aurora 7: The Mercury Space Flight of M. Scott Carpenter* (Chichester, 2016), pp. 101–24.

12 Walter Cronkite, live CBS television report, 24 May 1962.

13 Francis French and Colin Burgess, *Into That Silent Sea: Trailblazers of the Space Era, 1961–1965* (Lincoln, NE, 2007), p. 184.

14 'U.S. Stunned by Soviet Double', *Daily Mirror* (Sydney), 13 August 1962, p. 4.

15 Evgeny Riabchikov, *Russians in Space* (New York, 1971), p. 190.

16 Sobel, ed., *Space*, pp. 169–70.

17 'Schirra's Space Thrill: Controls Cut Off in Orbit', *Daily Telegraph* (Sydney), 5 October 1962, p. 2.

18 Colin Burgess, *Faith 7: L. Gordon Cooper, Jr., and the Final Mercury Mission* (Chichester, 2016), pp. 88–122.

19 Gordon Cooper, 'Astronaut's Summary Flight Report', in *Project Mercury Summary, Including Results of the Fourth Manned Orbital Flight*, NASA SP-45 (Washington, DC, October 1963), p. 358.

20 Bart Hendrickx, 'The Kamanin Diaries 1960–1963', *Journal of the British Interplanetary Society*, L (1997), pp. 33–40.

21 David J. Shayler and Ian Moule, *Women in Space: Following Valentina* (Chichester, 2005), pp. 46–50.

22 French and Burgess, *Into That Silent Sea*, pp. 312–31.

## 4  STEPPING INTO THE VOID

1 Francis French and Colin Burgess, *Into That Silent Sea: Trailblazers of the Space Era, 1961–1965* (Lincoln, NE, 2007), p. 344.

2 Rex Hall and David J. Shayler, *The Rocket Men: Vostok and Voskhod, the First Soviet Manned Spaceships* (Chichester, 2001), pp. 233–4.

3 David J. Shayler and Colin Burgess, *NASA's Scientist-Astronauts* (Chichester, 2007), p. 52.

4 Lester A. Sobel, ed., *Space: From Sputnik to Gemini* (New York, 1965), p. 270.

5 Hall and Shayler, *The Rocket Men*, p. 246.

6 David Scott and Alexei Leonov, *Two Sides of the Moon* (London, 2004), p. 109.

7 Ibid.

8 French and Burgess, *Into That Silent Sea*, pp. 363–4.

9 Asif Siddiqi, 'Cancelled Missions in the Voskhod Program', *Journal of the British Interplanetary Society*, L/1 (January 1997), pp. 25–31.

10 NASA Manned Spacecraft Center, Houston, Texas, News Release: *Manned Space Flight Comes of Age as Project Mercury Nears Its End*, January 1962, p. 3.

11 'Naming Mercury-Mark II Project', memorandum from D. Brainerd Holmes, director of Manned Space Flight Programs, to Associate Administrator, NASA, 16 December 1961, NASA History Division, Folder 18674.

12 Colin Burgess, *Liberty Bell 7: The Suborbital Flight of Virgil I. Grissom* (Chichester, 2014), pp. 206–7.

13 Sobel, ed., *Space*, p. 274.

14 'Rocket Casing Missed by Spacemen', *The Sun* (Sydney), 5 June 1965, pp. 2, 4.

15 Francis French and Colin Burgess, *In the Shadow of the Moon: A Challenging Journey to Tranquility, 1965–1969* (Lincoln, NE, 2007), p. 32.

16 'Chasing a Space Case', *The Australian*, 20 August 1965, p. 5.

17 Colin Burgess, *Faith 7: L. Gordon Cooper, Jr., and the Final Mercury Mission* (Chichester, 2016), p. 222.

18 Colin Burgess, *Sigma 7: The Six Mercury Orbits of Walter M. Schirra, Jr.* (Chichester, 2016), p. 226.

19 David Scott and Alexei Leonov, *Two Sides of the Moon: Our Story of the Cold War Space Race* (New York, 2004), p. 101.

20 Flora Lewis, 'Gemini-9's Lost Bird Slows U.S. Moon Race', *The Australian*, 18 May 1966, p. 8.

21 'Space Walk for Repairs: Satellite Crippled', *Daily Mirror* (Sydney), 4 June 1966, p. 1.

22 Eugene Cernan with Don Davis, *The Last Man on the Moon* (New York, 1999), p. 135.

23 Dr David R. Williams, 'The Gemini Program (1962–1966)', https://nssdc.gsfc.nasa.gov, accessed 15 November 2020.

### 5 TRAGEDY ON THE LAUNCH PAD

1 Donald K. Slayton with Michael Cassutt, *Deke: U.S. Manned Space from Mercury to the Shuttle* (New York, 1994), p. 164.

2 Francis French and Colin Burgess, *In the Shadow of the Moon: A Challenging Journey to Tranquility, 1965–1969* (Lincoln, NE, 2007), p. 200.

3 Colin Burgess, *Liberty Bell 7: The Suborbital Mercury Flight of Virgil I. Grissom* (Chichester, 2014), p. 222.

4 Colin Burgess and Kate Doolan with Bert Vis, *Fallen Astronauts: Heroes Who Died Reaching for the Moon* (Lincoln, NE, 2016), p. 118.

5 George Leopold, *Calculated Risk: The Supersonic Life and Times of Gus Grissom* (West Lafayette, IN, 2016), p. 222.

6 Burgess and Doolan with Vis, *Fallen Astronauts*, p. 128.

7 William Harwood, 'Apollo 1 Crew Honored 50 Years after Fatal Fire', www.cbsnews.com, 27 January 2017.

8 Report: *Apollo 204 Review Board to the Administrator National Aeronautics and Space Administration, Appendix D*, tabled 5 April 1967, available at https://history.nasa.gov

9 Walter Cunningham, *The All-American Boys* (New York, 1977), p. 15.

10 Richard Hollingham, 'The Fire That May Have Saved the Apollo Programme', www.bbc.com, 27 January 2017.

11 Colin Burgess, *Sigma 7: The Six Mercury Orbits of Walter M. Schirra, Jr.* (Chichester, 2016), p. 235.

12 Ibid., p. 237.

13 'Astronaut Wally Schirra, 84, Dies', www.tulsaworld.com (via Associated Press Wire Services), 4 May 2007.

14 Jamie Doran and Piers Bizony, *Starman: The Truth behind the Legend of Yuri Gagarin* (London, 1998), p. 187.

15 Burgess and Doolan with Vis, *Fallen Astronauts*, pp. 243–5.
16 Alexander Petrushenko (Moscow), correspondence with the author, June 1992.

## 6 EYES ON THE MOON

1 Melanie Whiting, ed., *50 Years Ago: Considered Changes to Apollo 8*, NASA History Office, www.nasa.gov, 9 August 2018.
2 Robert Kurson, *The Daring Odyssey of Apollo 8 and the Astronauts Who Made Man's First Journey to the Moon* (New York, 2018), p. 244.
3 Christian Davenport, 'Earthrise: The Stunning Photo That Changed How We See Our Planet', *Washington Post*, 24 December 2018.
4 *Soviet Space Programs, 1962–65: Goals and Purposes, Achievements, Plans, and International Implications*, prepared by the Committee on Aeronautical and Space Sciences, U.S. Senate, 89th Congress, 2nd session (Washington, DC, December 1966), pp. 388–9.
5 David Scott and Alexei Leonov with Christine Toomey, *Two Sides of the Moon* (London, 2004), p. 239.
6 Ibid.
7 Ibid.
8 Becky Little, 'The Soviet Response to the Moon Landing? Denial There Was a Moon Race at All', www.history.com/news, 11 July 2019.
9 Kelli Mars, ed., *50 Years Ago, Apollo 8 Is GO for the Moon*, NASA History Office, Apollo 8, www.nasa.gov, 14 November 2018.
10 Donald K. Slayton with Michael Cassutt, *Deke: U.S. Manned Space from Mercury to the Shuttle* (New York, 1994), p. 207.
11 Scott and Leonov with Toomey, *Two Sides of the Moon* (London, 2004), p. 237.
12 Eugene Cernan with Don Davis, *The Last Man on the Moon* (New York, 1999), p. 215.

## 7 ONE GIANT LEAP

1 *Roundup* (NASA JSC Space Center magazine), VIII/7 (24 January 1969), p. 1.
2 James R. Hansen, *First Man: The Life of Neil A. Armstrong* (New York, 2005).
3 Douglas B. Hawthorne, *Men and Women of Space* (San Diego, CA, 1992), pp. 33–5.
4 Francis French and Colin Burgess, *In the Shadow of the Moon* (Lincoln, NE, 2007), pp. 123–6.
5 Hawthorne, *Men and Women of Space*, pp. 154–7.
6 John M. Mansfield *Man on the Moon*, 1st edn (Sheridan, OR, 1969), pp. 204–5.
7 Neil Armstrong, Michael Collins and Edwin E. Aldrin Jr, *First on the Moon*, 1st edn (Toronto and Boston, MA, 1970), p. 289.
8 Mansfield, *Man on the Moon*, p. 226.
9 NASA Marshall Space Flight Center report, *Analysis of Apollo 12 Lightning Incident*, MSC-01540, available at https://spaceflight.nasa.gov, February 1970.

10 Rick Houston and Milt Heflin, *Go Flight: The Unsung Heroes of Mission Control, 1965–1992* (Lincoln, NE, 2015), pp. 184–6.

11 Gene Kranz, 'Apollo 13', talk at National Air and Space Museum, Washington, DC, 8 April 2005.

12 Tim Furniss and David J. Shayler with Michael D. Shayler, *Praxis Manned Spaceflight Log, 1961–2006* (Chichester, 2007), p. 138.

13 Kranz, 'Apollo 13'.

14 Donald K. Slayton with Michael Cassutt, *Deke! U.S. Manned Space from Mercury to the Shuttle*, 1st edn (New York, 1994), p. 258.

15 Ibid.

16 Kranz, 'Apollo 13'.

17 Lunar and Planetary Institute, 'Apollo 14 Mission Overview', www.lpi.usra.edu/lunar/missions/apollo/apollo_14, 2019.

18 Al Worden with Francis French, *Falling to Earth: An Apollo 15 Astronaut's Journey to the Moon* (Washington, DC, 2011), pp. 213–14.

19 Furniss and D. Shayler with M. Shayler, *Praxis Manned Spaceflight Log*, pp. 158–9.

20 Eugene Cernan and Donald A. Davis, *The Last Man on the Moon* (New York, 2007), p. 337.

21 Colin Burgess, *Shattered Dreams: The Lost and Canceled Space Missions* (Lincoln, NE, 2019), pp. 28–31.

22 'Peace Call as Moon Mission Ends', *Canberra Times*, 15 December 1972, pp. 1, 5.

23 Eileen Stansbury, 'Lunar Rocks and Soils from Apollo Missions', https://curator.jsc.nasa.gov, 1 September 2016.

## 8 SOVIET SETBACK AND SKYLAB

1 Colin Burgess and Rex Hall, *The First Soviet Cosmonaut Team: Their Lives, Legacy and Historical Impact* (Chichester, 2009), pp. 285–7.

2 NASA Scientific and Technical Information Office, *Astronautics and Aeronautics, 1971: Chronology on Science, Technology and Policy*, NASA publication SP–4016 (Washington, DC, 1973), p. 105.

3 Rex D. Hall and David J. Shayler, *Soyuz: A Universal Spacecraft* (New York, 2003), pp. 174–5.

4 Colin Burgess and Kate Doolan with Bert Vis, *Fallen Astronauts: Heroes who Died Reaching for the Moon* (Lincoln, NE, 2016), pp. 260–61.

5 Derryn Hinch, 'Prisoners of the Earth?', *Sydney Morning Herald*, 1 July 1971, p. 7.

6 'Bubbles in Blood Killed 3 Soviet Spacemen', *Sydney Daily Mirror*, 2 July 1971, p. 3.

7 Burgess and Doolan with Vis, *Fallen Astronauts*, p. 262.

8 Ben Evans, 'The Plan to Save Skylab (Pt 2)', *Space Safety Magazine*, 20 May 2013, www.spacesafetymagazine.com.

9 'Tired Nauts Link to Skylab', *San Francisco Chronicle*, 26 May 1973, p. 1.

10 Derryn Hinch, 'Skylab Space Station Gets a Sunshield', *Sydney Morning Herald*, 27 May 1973, p. 4.

11 Alex Faulkner, 'Skylab Fine as Crew Move In', *Daily Telegraph*, 27 May 1973, p. 3.

12 'Skylab-2: Mission Accomplished!', www.nasa.gov/feature/skylab-2-mission-accomplished, 22 June 2018.

13 'Second Crew Join Skylab in Orbit: Astronauts Set for 59 Days in Space', *Sydney Morning Herald*, 30 July 1973, p. 6.

14 Kathleen Maughan Lind, *Don Lind: Mormon Astronaut* (Salt Lake City, UT, 1985), pp. 143–4.

15 'Skylab 2's Space Leak Is Solved', *Daily Mirror* (Sydney), 6 August 1973, p. 3.

16 'Houston Chides Skylab Crew for Hiding Pogue's Vomiting', *International Herald Tribune*, 19 November 1973, p. 1.

17 'Kohoutek Turns Mysterious', *Straits Times* (Singapore), 30 December 1973, p. 7.

18 David Shayler, *Around the World in 84 Days: The Authorized Biography of Skylab Astronaut Jerry Carr* (Burlington, ON, 2008), pp. 203–4.

19 Don L. Lind interviewed by Rebecca Wright, Houston, Texas, 27 May 2005, NASA JSC Oral History Project, https://historycollection.jsc.nasa.gov.

## 9 RECOVERING THE SOYUZ/SALYUT MISSIONS

1 *Soviet Space Programs, 1976–1980 (Part 2)*, Report prepared for the 98th Congress, 2nd session, United States Senate, for the Committee on Commerce, Science and Transportation, October 1984, p. 548, available online at https://files.eric.ed.gov.

2 'Mission Misfire', *Time*, CV/16 (21 April 1975), p. 38.

3 Geoffrey Bowman, 'The Last Apollo', in *Footprints in the Dust: The Epic Voyages of Apollo, 1969–1975*, ed. Colin Burgess (Lincoln, NE, 2010), pp. 386–8.

4 Paul Recer (Associated Press), 'They Shake Hands in Space', *San Francisco Examiner*, 17 July 1975, p. 1.

5 Tim Furniss and David J. Shayler with Michael D. Shayler, *Praxis Manned Spaceflight Log, 1961–2006* (Chichester, 2007), p. 193.

6 'Russians in Second Space Docking', *The Australian*, 13 January 1977, p. 3.

7 Furniss and D. Shayler with M. Shayler, *Praxis Manned Spaceflight Log*, p. 262.

8 Umberto Cavallaro, *Women Spacefarers: Sixty Different Paths to Space* (Cham, 2017), p. 14.

9 'NASA to Recruit Space Shuttle Astronauts', 8 July 1976, NASA news release 76-44, Johnson Space Center, Houston, TX.

10 David J. Shayler and Colin Burgess, *NASA's First Space Shuttle Astronauts: Redefining the Right Stuff* (Cham, 2020), pp. 17–18.

11 Colin Burgess and Bert Vis, *Interkosmos: The Eastern Bloc's Early Space Program* (Cham, 2016).

12 Furniss and D. Shayler with M. Shayler, *Praxis Manned Spaceflight Log*, pp. 345–6.

13 Eric Betz, 'The Last Soviet Citizen', *Discover*, 1 December 2016, www.discovermagazine.com.

14 Clay Morgan, 'Jerry Linenger: Fire and Controversy, January 12–May 24, 1997', chapter in *Shuttle–Mir: The United States and Russia Share History's Highest Stage*, NASA JSC History Series, NASA Publication SP-4225, (Washington, DC, 2011), p. 92.

15 Ibid., p. 109.


## 10 SPACE SHUTTLES AND THE ISS

1 John M. Logsdon, *Ronald Reagan and the Space Frontier* (Cham, 2018), p. 35.

2 Lynn Sherr, *Sally Ride: America's First Woman in Space* (New York, 2014), pp. 159–65.

3 Tim Furniss and David J. Shayler, *Praxis Manned Spaceflight Log, 1961–2006* (Chichester, 2007), pp. 278–80.

4 Kelli Mars, ed., '35 Years Ago, STS-9: The First Spacelab Science Mission', NASA History Office, www.nasa.gov, 28 November 2018.

5 Colin Burgess, *Teacher in Space: Christa McAuliffe and the Challenger Legacy* (Lincoln, NE, 2000), pp. 101–2.

6 Tim Furniss, 'Shuttle Leaves Leasat Adrift', *Flight International*, 27 April 1986, p. 18.

7 'Shuttle Mission Success', *Flight International*, 28 June 1986, p. 6.

8 Furniss and Shayler, *Praxis Manned Spaceflight Log*, pp. 324–5.

9 Jeanne Ryba, ed., 'NASA Mission Archives: STS-61A', NASA John F. Kennedy Space Center, www.nasa.gov, updated 18 February 2010.

10 Colin Burgess, *Teacher in Space: Christa McAuliffe and the Challenger Legacy* (Lincoln, NE, 2000), pp. 76–80.


## 11 EXPANDING THE SPACE FRONTIER

1 Lee Dye, 'American Back in Space with Majestic Launch of *Discovery*', *Los Angeles Times*, 30 September 1988, p. 1.

2 '*Discovery*'s "Great Ending"', *Los Angeles Daily News*, 4 October 1988, p. 1.

3 'The First Orbiting Solar Observatory', NASA Goddard Space Flight Center, www.gsfc.nasa.gov, 26 June 2003.

4 Brian Dunbar, 'Hubble's Mirror Flaw', NASA Media Resources, www.nasa. gov, updated 26 November 2019.

5 Columbia Accident Investigation Board Report, vol. 1, part 2, Chapter 5, 'Why the Accident Occurred', p. 97.

6 Columbia Accident Investigation Board Report, vol. 1, part 1, Chapter 11, 'Return to Flight Recommendations', p. 225.

7 Miles O'Brien for CNN.com (International), 'NASA Chief to Resign', http://edition.cnn.com, 13 December 2004.

8 Colin Burgess, 'The Final Countdown', *Australian Sky and Telescope* (May 2005), pp. 35–8.

9 'The Fight to Save Skylab', *Flight International*, 24 May 1973, pp. 810–11.

10 NASA *Space Station Freedom Media Handbook* (NASA Archive Document NASA-TM-10291), Washington, DC, April 1989, pp. 4–6, available online at https://ntrs.nasa.gov/citations/19900014144.

11 Tim Furniss and David J. Shayler, *Praxis Manned Spaceflight Log, 1961–2006* (Chichester, 2007), pp. 665–7.

12 Chris Bergin, 'Remembering Buran – The Shuttle's Estranged Soviet Cousin', www.nasaspaceflight.com, 15 November 2013.

13 Marina Koren, 'China's Growing Ambitions in Space', *The Atlantic*, www.theatlantic.com, 23 January 2017.

14 Matthew S. Williams, 'All You Need to Know about China's Space Program', https://interestingengineering.com, 16 March 2019.

15 Michael Cassutt, 'Citizen in Space', *Space Illustrated* (February 2001), pp. 27–9.

16 Colin Burgess, 'All Systems Go!', *Australian Sky and Telescope* (November 2005), pp. 25–9.

17 Jonathan Amos, 'Sarah Brightman Calls Off Space Trip', www.bbc.com, 14 May 2015.

18 Sandhya Ramesh, 'India Says It Will Send a Human to Space by 2022', the Planetary Society, www.planetary.org, 24 August 2018.

19 'Four IAF Men to Train as Astronauts for Gaganyaan Mission: ISRO', *New Indian Express*, www.newindianexpress.com, 1 January 2020.

## EPILOGUE: OUR FUTURE IN SPACE

1 Rebecca Anderson and Michael Peacock, *Ansari X-Prize: A Brief History and Background*, NASA History, www.history.nasa.gov, updated 5 February 2010.

2 Leonard David, 'SpaceShipOne Wins $10 million Ansari X-Prize in Historic 2nd Trip into Space', www.space.com, 4 October 2004.

3 Natasha Bernal, 'Sir Richard Branson's Space Race: Over Two Decades of Broken Promises', *Daily Telegraph*, 9 July 2019, p. 16.

4 Mahita Gajanan, 'Virgin Galactic Crash: Co-pilot Unlocked Braking System, Enquiry Finds', www.theguardian.com, 29 July 2015.

5 Erik Seedhouse, 'Space Tourism', www.britannica.com, accessed 17 November 2020.

6 Alan Boyle, 'Space Racers Unite in Federation', www.nbcnews.com, 2 August 2005.

7 Peter N. Spotts, 'Private Space Tourism Takes Off', www.christiansciencemonitor.com, 21 July 2005.

8 Virgin GalacticPress, 'Virgin Galactic and Social Capital Hedorophia Announce Merger', www.virgingalactic.com, 9 July 2019.

9 Erin Mahoney, ed., 'Q&A: NASA's New Spaceship', NASA Johnson Space Center, online at www.nasa.gov, 13 November 2018.

10 Christian Davenport, *The Space Barons: Elon Musk, Jeff Bezos, and the Quest to Conquer the Cosmos* (New York, 2018).

# SELECT BIBLIOGRAPHY

Baker, David, *The History of Manned Space Flight* (New York, 1981)

Brzezinski, Matthew, *Red Moon Rising: Sputnik and the Hidden Rivalries That Ignited the Space Age* (New York, 2007)

Burgess, Colin, and Kate Doolan, *Fallen Astronauts: Heroes Who Died Reaching for the Moon* (Lincoln, NE, 2016)

—, and Chris Dubbs, *Animals in Space: From Research Rockets to the Space Shuttle* (New York, 2007)

—, and Rex Hall, *The First Soviet Cosmonaut Team: Their Lives, Legacy and Historical Impact* (New York, 2009)

—, and Bert Vis, *Interkosmos: The Eastern Bloc's Early Space Program* (New York, 2016)

Carpenter, M. Scott, L. Gordon Cooper Jr, John H. Glenn Jr, Virgil I. Grissom, Walter M. Schirra Jr, Alan B. Shepard Jr and Donald K. Slayton, *We Seven: By the Astronauts Themselves* (New York, 1962)

Cernan, Eugene, and Don Davis, *The Last Man on the Moon* (New York, 1999)

Chaikin, Andrew, *A Man on the Moon: The Voyages of the Apollo Astronauts* (New York, 1994)

Dickson, Paul, *Sputnik: The Shock of the Century* (New York, 2001)

Doran, Jamie, and Piers Bizony, *Starman: The Truth behind the Legend of Yuri Gagarin* (London, 1998)

Dubbs, Chris, and Emeline Paat-Dahlstrom, *Realizing Tomorrow: The Path to Private Spaceflight* (Lincoln, NE, 2011)

Evans, Michelle, *The X-15 Rocket Plane: Flying the First Wings in Space* (Lincoln, NE, 2013)

French, Francis, and —, *In the Shadow of the Moon: A Challenging Journey to Tranquility, 1965–1969* (Lincoln, NE, 2007)

—, and —, *Into That Silent Sea: Trailblazers of the Space Era, 1961–1965* (Lincoln, NE, 2007)

Furniss, Tim, David J. Shayler and Michael D. Shayler, *Praxis Manned Spaceflight Log, 1961–2006* (Chichester, 2007)

Hall, Rex D., and David J. Shayler, *Soyuz: A Universal Spacecraft* (New York, 2003)

—, —, and Bert Vis, *Russia's Cosmonauts: Inside the Yuri Gagarin Training Center* (New York, 2005)

Hitt, David, Owen Garriott and Joe Kerwin, *Homesteading Space: The Skylab Story* (Lincoln, NE, 2008)

Kluger, Jeffrey, *Apollo 8: The Thrilling Story of the First Mission to the Moon* (New York, 2017)

Murray, Charles, and Catherine Bly Cox, *Apollo: The Race to the Moon* (New York, 1989)

Pyle, Rod, *Space 2.0: How Private Spaceflight, a Resurgent NASA, and International Partners are Creating a New Space Age* (Dallas, TX, 2019)

Riabchikov, Evgeny, *Russians in Space*, trans. Guy Daniels (Garden City, NY, 1971)

Shayler, David J., and Michael D. Shayler, *Manned Spaceflight Log II, 2006–2012* (New York, 2013)

Shelton, William, *Soviet Space Exploration: The First Decade* (London, 1968)

Slayton, Donald K., and Michael Cassutt, *Deke! U.S. Manned Space: From Mercury to the Shuttle* (New York, 1994)

Wolfe, Tom, *The Right Stuff* (New York, 1979)

Worden, Al, and Francis French, *Falling to Earth: An Apollo 15 Astronaut's Journey to the Moon* (Washington, DC, 2011)

# ACKNOWLEDGEMENTS

As usual, I am deeply indebted to many of my spaceflight colleagues, my family and friends – both here in Australia and overseas – for their time, expertise and advice. Due to the COVID-19 pandemic, it has been an unusual and troubling time for many, with enforced relegation to isolation and home computers becoming almost the norm throughout the publishing industry. So, for their patience and understanding, I would like to offer my first round of thanks to the amazing and talented team at Reaktion Books: Michael Leaman, Alex Ciobanu, Susannah Jayes and Amy Salter, and anyone else who may have been anonymously (to me) involved in the editing and preparation of this book.

Many thanks also to Peter Morris, Emeritus Senior Research Fellow at the Science Museum in London, who, as front-line editor for Reaktion Books, first touched base with me and asked if I might be interested in putting this book together. It has proved to be a wonderful, sometimes frustrating, but above all exciting collaboration. Thanks also to Peter's colleague Doug Millard, Deputy Keeper, Technologies and Engineering at the Science Museum.

Suitable and good-quality photographs of early human space endeavours are often hard to locate, and I am therefore grateful for the kind and continuing help of Joachim Becker at Spacefacts.de and Ed Hengeveld. Their assistance has lent much to the illustrative excellence of this publication.

Finally, to the legion of people and friends who have helped me collate my spaceflight files over many decades, my heartfelt thanks. Although far too numerous to mention by name, there has always been a solid vanguard of helpers ready to assist me in my various efforts, such as David Shayler, Bert Vis, Michael Cassutt, Francis French and the late Rex Hall MBE.

This book is hopefully a worthy recognition of the international teamwork involved in its research, editing and production during a particularly difficult time in human history.

# PHOTO ACKNOWLEDGEMENTS

The author and publishers wish to express their thanks to the below sources of illustrative material and/or permission to reproduce it.

Author's collection: pp. 23, 46, 58, 60, 80, 94, 100, 101, 106, 109, 110 (illustration by Alexei Leonov), 111, 159, 160, 244, 317; courtesy of Joachim Becker, Spacefacts. de: pp. 162, 220, 222, 225, 252, 256, 316, 319, 312 centre; Blue Origin: pp. 338 bottom, 339; courtesy of Celestia free software: p. 87; Keith McNeill, Space Models Photography: p. 321 top; NASA: pp. 37, 40, 42, 50, 55, 64, 67, 70, 83, 91, 95, 97, 117, 118, 120, 123, 124, 127 top and bottom, 128, 129, 131, 133, 135, 137, 138, 139, 143, 150, 152, 153, 167, 168, 175, 177, 180, 182, 191, 193, 194, 196, 198, 200, 203, 206, 208, 209, 211, 212, 214, 216, 217, 228, 229, 232, 234, 235, 237, 239, 246, 247, 248, 249, 253, 254, 260, 262, 266, 269, 272, 275, 277, 281, 282, 285, 290, 291, 293, 294, 302, 305, 306, 307, 308, 309, 312, 314, 323, 337 bottom, 340; NASA/Associated Press: p. 15; NASM Archives: p. 43; NORAD and USNORTHCOM Public Affairs: pp. 334 top, 336; Scaled Composites, LLC: pp. 329, 333 (via Wikimedia Commons); courtesy of Dr David Simons: p. 32; photograph by Robert Sisson, with permission of National Geographic Society: p. 34; SpaceX: pp. 334 top and centre, 337 top, 338 centre; SpaceX/NASA: p. 335; USAF: p. 74; U.S. Army: pp. 10, 19, 33; courtesy of Virgin Galactic: pp. 328, 330; Wikimedia Commons: p. 20.

# INDEX

Page numbers in *italics* indicate illustrations